DESIGN and USE of RELATIONAL DATABASES in CHEMISTRY

DESIGN and USE of RELATIONAL DATABASES in CHEMISTRY

TJ O'Donnell

CRC Press
Taylor & Francis Group
Boca Raton   London   New York

CRC Press is an imprint of the
Taylor & Francis Group, an **informa** business

CRC Press
Taylor & Francis Group
6000 Broken Sound Parkway NW, Suite 300
Boca Raton, FL 33487-2742

© 2009 by Taylor & Francis Group, LLC
CRC Press is an imprint of Taylor & Francis Group, an Informa business

Library of Congress Cataloging-in-Publication Data

O'Donnell, T. J.
    Design and use of relational databases in chemistry / T.J. O'Donnell.
        p. cm.
    Includes bibliographical references and index.
    ISBN 978-1-4200-6442-1 (hardcover : alk. paper)
    1. Chemistry--Databases. 2. Chemistry--Data processing. 3. Relational databases. I. Title.

QD455.3.E4O46 2009
542'.85--dc22   2008040765

Visit the Taylor & Francis Web site at
http://www.taylorandfrancis.com

and the CRC Press Web site at
http://www.crcpress.com

Contents

Preface

When I first encountered databases, I thought of them as simple repositories of large amounts of data. Anytime a project required data stored there, I exported what I needed into flat files for further work. After all, most computational chemistry tools available then were designed to read and write files. Sometimes the database contained structured information that was not easy to represent in flat files, so I created computer program objects and data structures to do so. For example, in a project requiring a web interface for users of a chemical stockroom, I duplicated the relationship among compounds, samples, cabinet, and shelf locations in data structures in my perl programs. Instead of using methods that are built into a relational database system, I imported the entire set of tables into my program in order to work with them. After learning enough about relational databases to read complex sets of tables, I came to see the benefits of having data organized in relational tables. I also came to see the benefit of using the database to access only the required data and only when necessary.

Today, I have turned my habit around. When I have a set of chemical structures or data files, my first task is to organize them in a relational database. After all, the tools I now use are designed to read and write tables in a database. Rather than creating folders to keep project files, I create a schema of tables with rows holding chemical structures and data imported from the files. For example, the PubChem project provides information on millions of compounds in the form of hundreds of chemical structure files and associated experimental data files. While PubChem provides excellent Web tools to search this data, for local use I developed a schema to hold the structures and data in related tables. One possible schema for this is shown in Chapter 6 of this book.

Relational databases are not just for storage and organization of information. While they have always provided useful tools to search data, I now realize how the extensibility of a database increases its usefulness by using various procedural languages operating within the database. I have converted most of my essential tools into database functions. Whenever possible, these tools operate on whole tables rather than processing them using arrays and iterators. It seems now that I am inside the database looking out, rather than the other way around. Of course, there will always be a need for data processing outside the database, especially for programs that require hours, day, or weeks to complete their work. However, most of my everyday work with chemical information benefits greatly from the organization, data integrity, and extensibility inherent in a relational database.

Acknowledgments

The work in this book would not have been possible without the support of many people and organizations. The PostgreSQL community has created and maintains the world's most advanced open-source relational database. Matthew Stahl and OpenEye Scientific Software have supported my work in chemical relational databases by providing me access to their OEChem library. Geoffrey Hutchison and Noel O'Boyle helped me with OpenBabel, which has benefited from the contributions of Craig James. I thank Ivan Tubert-Brohman for PerlMol, Brian Kelley for FROWNS, and Andrew Dalke for discussions on FROWNS. Dave Weininger, Daylight Chemical Information Systems, Inc., their annual MUG meetings, and Jack Delany's early work on chemical data cartridges have been inspiring. I thank Tom Doman for years of collaborative work. He and Kerry Fowler have encouraged me and helped me with discussions about topics in this book and about other matters, chemical and not. Many good suggestions and feedback on the book were offered by Maya Binun, David Goodsell, and Robin Hewitt, who still laughs at how enthusiastically I have adopted relational databases. Mike Richards has been a great help with SQL questions. He and Joseph Cohen at Senomyx make better use of relational databases than anyone I know. I am indebted to all my bosses and mentors over the years, including Pierre LeBreton, Tom DeFanti, Art Olson, Yvonne Martin, and Jim Snyder. I thank all of my colleagues and clients over the years who have come to me with interesting projects that stretched my limits. The independent coffee house owners and baristas of San Diego made my work on this book an enjoyable social and caffeinated experience.

Biography

TJ O'Donnell earned his Ph.D. from the University of Illinois at Chicago in 1980 for work involving quantum chemical computations and photoelectron spectroscopy. While working at Abbott Laboratories in the early 1980s, he developed one of the early molecular modeling systems that incorporated interactive three-dimensional computer graphics. Since 1987, he has worked as an independent consultant in the field of computational chemistry, primarily in drug design groups in the pharmaceutical industry. In 2005, he formed gNova, Inc. to include consulting and sales of chemical database software. He now resides in San Diego, California.

chapter 1

Introduction

The goal of this book is to convince you that relational databases are the best way to store, search, and even operate on chemical information. Whether the database contains a hundred structures or ten million, a relational database provides ways to ensure data integrity, to formalize relationships among the data, and to extend the database when new data become available or when new ways of operating on data become of interest.

Some readers of this book will have a background in chemistry and wish to see how databases might assist their work. Some readers will already have a background in programming and databases. After reading this book, you should have the ability to understand an existing relational database schema or design a new schema containing tables of data and chemical structures. You will learn how to take advantage of database extension "cartridges" that provide ways of properly storing and searching chemical structures, not just numerical or textual data. You will see how you can download and install a fully functioning database with free and open-source chemical extension cartridges. You will also see how the database can be accessed on a computer network using existing applications or ones that you wish to write.

There are many books that describe relational database management systems (RDBMS) and the structured query language (SQL) used to manipulate the data. Understanding SQL is important, and this book contains an introduction to SQL. However, the focus is on the concepts of relational data. One goal is to show how a proper integration of a new molecular structure data type yields a powerful, extended relational database for use in chemistry. For those of you new to relational databases, it is expected that the SQL introduction will suffice for your understanding of the concepts in this book. For those of you already familiar with SQL, it is hoped that you will see how the extensions described here provide a powerful, integrated way to handle molecular structures within the database. In either case, there are plenty of practical SQL examples contained in this book.

Much of this book is a discussion of computer languages. SQL is a type of programming language. Becoming fluent in SQL helps make the most of a relational database. SMILES, SMARTS, and SMIRKS are chemical computer languages that express many fundamental aspects of chemical structure. Becoming fluent in all these languages will help you create and maintain robust and powerful chemical relational databases. And all four languages begin with S! Finally, you will see how you can use your familiarity with perl, python, or C to implement new functions in the database.

Much chemical data is stored in computer files, some of which have little or no structural organization. Some data files are more structured, perhaps in tabular form or as an Excel spreadsheet. There are many similarities between spreadsheet files and relational tables in a database. However, storing data in a relational database offers many advantages not possible when data is stored in files. The greatest advantage comes from the proper design and use of tables themselves. Chapter 2 shows how to design and use tables to store and search numerical or text data. The reason for using multiple tables is explained and the use of relationships among tables is examined. Finally, the entity-relationship diagram is shown as an aid to designing and understanding a database of tables.

An introduction to SQL is provided in Chapter 3, but with an emphasis on examples relevant to chemical information rather than business information, which is often used in other books. Chapter 4 discusses some of the RDBMS that are available, namely Oracle, MySQL, and PostgreSQL. All of them use SQL to insert, delete, update, and select data. Chapter 5 shows ways in which client programs, including Web-based applications, are used to connect to the database server. Chapter 6 examines ways in which RDBMS are typically used to handle numerical and textual chemical information using relational tables. An example of using data files from the PubChem project is included.

Chapter 7 introduces ways in which RDBMS can be used to handle chemical structural information using SMILES and SMARTS representations. It shows how extensions to relational databases allow chemical structural information to be stored and searched efficiently. In this way, chemical structures themselves can be stored in data columns. Once chemical structures become proper data types, many search and computational options become available. Conversion between different chemical structure formats is also discussed, along with input and output of chemical structures.

Chapter 8 shows ways in which molecular fragments can be used to speed up searches for chemical structures. Both path-based and fragment-based methods are discussed. Several types of molecular similarity are explained using bit-string fingerprints representing the presence or absence of various fragments. Finally, it is shown how tables of fragments along with parameter values for these fragments can be used to compute theoretical molecular properties.

In Chapter 9, chemical reactions and transformations are discussed. Using SMIRKS to represent chemical transformations, reaction specifications can be stored in the database. Structures can be transformed and combined (reacted) to produce new structures.

New SQL functions and data types can be used to extend a relational database. This is explained in Chapter 10 using PostgreSQL as an example. Ways in which three-dimensional molecular structures can be stored are examined in Chapter 11. This chapter also advocates using an RDBMS instead of molecular structure files and shows how this transition might be accomplished.

Chapter 12 discusses more fully the ways in which client programs can interface with the database. The intent is to show how you might integrate a relational database into an existing suite of programs or design and implement a new computer system for chemical information.

Chapter 13 shows sample applications that might be developed to produce a registry of compounds for use within a company or project. A set of utility functions is discussed that allows molecular structure files to be imported into a database and used in various ways.

The Appendix to this book contains a complete set of SQL statements conforming to the examples used in the book. This book describes a core set of molecular structure functions and builds upon that. The methods used to build the core tools are completely contained in this book, mostly in the form of SQL functions and tables of data. Three different ways in which the core functionality can be implemented are shown in the Appendix. These are all based on the free and open-source PostgreSQL database and use free and open-source perl or python modules. There are also commercial chemical cartridge extensions to PostgreSQL and Oracle that may even more closely suit your needs. The methods in this book most closely parallel the methods available in the PostgreSQL extension cartridge CHORD from gNova, Inc. However, most of the methods described, except perhaps for those using bit strings (which are not supported in Oracle), could be implemented using any of the commercial toolkits or cartridges.

The Web site at http://www.gnova.com contains an implementation of every function in this book. There is also a database of structures and data. Using this resource, it is possible to try any of the techniques described in this book. Throughout the book, many practical examples of experimental, theoretical, and structural chemical data relations are described. The examples are available online, allowing you to connect to a live database and experiment with various search and display options.

chapter 2

Relational Database Fundamentals

2.1   Introduction

A relational database provides a way to store large amounts of data in tables that are either independent or related to one another. These tables have some similarities to spreadsheets, such as those used in Excel or OpenOffice. However, there are many advantages to storing data in relational tables. The purpose of this chapter is to provide guidelines for database design to ensure the creation of clear, extensible, and efficient database tables. There are many books with much more detailed information about database design, rules, and theory. After reading this chapter, you will be familiar with the concepts of tables, rows, columns, schemas, entity-relationship diagrams, primary keys, foreign keys, indexes, uniqueness, sequences, constraints, and joining tables.

The tables are formally called *relations,* referring to the mathematical set theory used in the original work on relational databases.[1] In database theory, rows are called *tuples* and columns are called *attributes of a tuple.* The focus of this book is practical, so the common terms *table, row,* and *column* are used. The detail of using the SQL language to perform these operations is left to a later chapter of this book.

2.2   Tables, Rows, and Columns

A table is a collection of data in rows and columns. As with tables in a scientific publication, each row typically represents some entity, such as a molecule, and each column represents some attribute of the entity, such as the name, molecular weight, ionization potential, or other theoretical or experimental data measurement. A table in a publication is laid out for clarity to the reader. Spreadsheet programs typically include ways to control the layout and look of the table. Display and layout features are irrelevant in a relational database.

A table in a relational database is intended to provide a consistent way to organize large amounts of data, constrain the data in meaningful ways, and extend the tables when new data becomes available. It does not contain any formatting or display information. Programs that access the database provide any display or formatting of the data in the table.

Table 2.1 Sample Table of Chemical Compound Data from EPA

Name	Formula	MW	logP	MP
Formaldehyde	CH_2O	30.03	0.35	−92
Guanidine hydrochloride	CH_6ClN_3	95.53	−3.56	182.3
Dexamethasone	$C_{22}H_{29}FO_5$	392.47	1.83	262
Cortisone acetate	$C_{23}H_{30}O_6$	402.49	2.1	222
Phenobarbital	$C_{12}H_{12}N_2O_3$	232.24	1.47	174
Oxyphenonium bromide	$C_{21}H_{34}BrNO_3$	428.41	0.17	191.5
Metharbital	$C_9H_{14}N_2O_3$	198.22	1.15	150.5
Mesantoin	$C_{12}H_{14}N_2O_2$	218.26	1.69	135
Meperidine	$C_{15}H_{22}ClNO_2$	283.80	3.03	187.5
Vitamin D2	$C_{28}H_{44}O$	396.66	10.44	116.5

Client programs are discussed in later chapters of this book. The structured query language (SQL) designed for creating, selecting, deleting, and updating the database is discussed in Chapter 3.

A relational table has a name, chosen when it is created. Although any name is possible, the name typically reflects the nature or source of the data contained in the table. Each column must also have a name. Consider Table 2.1, called EPA since it was constructed from data provided by the Environmental Protection Agency.[2] This table is readily understandable to any chemist. Each row contains information about one compound and each column contains a molecular attribute or property. In order to make it part of a relational database, a minimum of two things must be specified for each column: the column name and the column data type. In this example, the column names are Name, Formula, MW, logP, and MP corresponding to the compound name, molecular formula, molecular weight, octanol-water partition coefficient, and melting point. The column name in a relational table is arbitrary but is usually representative of the data contained in the column.

The nature of the data in each column must be specified by providing a data type. The data type must be one of a fixed set of types available in the relational database management system (RDBMS) being used. A discussion of several common RDBMS follows in Chapter 4. Some of the frequently used data types are

- Integer for whole numbers
- Numeric for possibly fractional numbers
- Text for character strings
- Date for dates
- Time for time-of-day values
- Timestamp for values containing both date and time

In the above example, Name and Formula are text and MW, logP, and MP are numeric. The order of the columns is fixed once it is specified when the table is created. The rows, however, are independent of each other.

There is no inherent internal order in which rows are stored in a table, regardless of the order in which they were inserted. When rows are selected, however, the order in which they are returned can be specified by sorting or other operations. If rows are selected without specific ordering instructions, the order is undefined and may change each time the rows are selected. An application must never rely on having the rows in a table returned in the same order, even if an ordering operation is performed. For example, it may be that the row for Phenobarbital is followed by the row for Phenobarbitone when the rows are sorted alphabetically by name. But if a row for Phenobarbitol is ever added to the table, it will appear after Phenobarbital when the selected rows are sorted by name.

2.3 External and Internal Representations of Data

A data type is necessary to allow the RDBMS to accurately convert the data from an external representation, most often text in a file, to an internal representation of the data. For example, the external representation of a numeric value is a text string containing at least one numeral, and possibly a plus or minus sign or a decimal point. A text value may contain any valid text character, usually only printable characters from the ASCII set. The internal representation of the data is dependent upon the particular RDBMS and hardware being used. It is not necessary to know the exact internal representation of the data. The important thing to consider is which data type accurately represents the data for your purposes.

There are rules governing the conversion from the external representation. These prevent improper data from being stored in the database. This is an important advantage over data stored in a typical computer file. It makes it impossible to store a nonnumeric value in a numeric column. For example, if a column is defined as numeric, an error would occur if the string ">2.97" were attempted to be inserted into that column of the table. Another advantage of enforcing these rules is that operations on the data in the columns can rely on the correct type of the data, not having to take special measures to handle nonconforming values. This applies to operations using SQL or operations performed by an external program that has selected data from the table. This data type concept will become especially important when new molecule data types are introduced in Chapter 7.

The internal representation of data is not entirely unimportant. When the float data type is used, the data are typically converted to the internal floating-point representation used by the computer on which the RDBMS is installed. This may have unintended consequences because of the rounding that occurs, especially if several mathematical operations are performed on the data in these columns. While it may be more computationally efficient to use the float data type and it may take less space in the database, the numeric data type is recommended for most scientific data, unless rounding errors are of no importance. There is also a double data type that lessens any rounding errors, but it does not prevent them.

Sometimes, there is no value available for a particular column or a particular row. Rather than inventing a special value to represent this, such as 99999 for numeric or "" for text, the relational database provides a special null value. This should be used when a value is unknown or unavailable. When actual data becomes available, the null value can be updated.

2.4 Advantages over Spreadsheets

The advantages of using data types discussed above apply to all relational tables regardless of size or complexity, but they are missing from spreadsheet programs, or are difficult to implement. Spreadsheet programs allow the data in a column to be formatted according to rules, such as for dates or numbers. However, these rules are simply formatting rules and are not enforced for all data input into a particular column. This reflects that the emphasis in typical spreadsheet programs on display rather than data. There are other valuable checks (see Constraints below) on data correctness that are easily implemented in relational tables but are impossible or clumsy when using spreadsheets.

2.4.1 Size and Speed

Table 2.1 could be easily and efficiently stored as a spreadsheet file. If the table grows to millions of rows, the ability of spreadsheet programs to update, sort, and otherwise manipulate the data would be severely impacted, or even impossible. A relational database has no inherent limit on the number of rows a table may contain because its rows are independent of one another. Even in a relational database, selecting or searching a table containing millions of rows takes longer than searching a table containing hundreds of rows, but the operation scales predictably. For example, if every row must be searched to select the ones desired, the operation will scale linearly.

In a relational database, data in a column may be indexed. This is explained in a later section of this chapter. If the relational table contains an index, many operations on the table can be greatly accelerated. Indexing is not possible in most spreadsheet programs.

2.4.2 Multiple Users

Only one person at a time can effectively use a spreadsheet, especially if it is being updated. Relational databases are designed to be used by multiple simultaneous users, even when tables are being updated. When necessary, particular tables or rows can be locked temporarily while being updated to prevent any accidental overwriting by two users. There are also many ownership and privilege options available in an RDBMS. These allow only some users to select data or to update data in the tables. These features are absent from spreadsheet programs.

2.5 Relationships among Tables

In a spreadsheet program, each table is essentially independent from other tables. This encourages the user to grow one table by tacking on new data columns as new data becomes available. This approach is also possible in a relational database and might even be more efficient than using multiple tables. However, it is better to organize information in separate tables and define relationships among the tables. A set of tables, with its associated rows and columns along with a definition of the relationships among them, is called a schema.

2.5.1 One-to-Many Relationships

Consider the EPA Table 2.1. There is no need to use separate tables to store this information. But, suppose that the need to store data on the water-octanol partition coefficient (logP) grew. For example, multiple measurements of logP might become important. These could be values measured at different temperatures or theoretical estimates of logP. It might be tempting to add other columns, named, say, logP1, Temp1, logP2, Temp2, clogP1, clogP2, and so forth. When additional columns such as these are added to a table, the table is said to violate normal form. Normal form is discussed more fully later in this chapter. It is better to create a new table to contain only the logP data.

This new logP table could in principle contain columns Temp1, logP2, Temp2, logP2, clogP1, clogP2, and so forth, but this is still not "the relational way" to store the data. Instead, consider the nature of the information to be stored in order to define which column the table will contain. Of course, the logP value itself is essential and must be one of the columns. The temperature is another important piece of information. Finally, the method used to measure or compute the value must be recorded. So the logP table would consist of three columns, logP as a numeric value, temperature as a numeric value, and method as text. Using these three columns and multiple rows, it is possible to store any number of values for logP along with the temperature and method. Notice that there may be multiple rows for any one compound.

The original table can now have the logP column removed, but how will the data in the logP table stay associated with the proper rows of the

Table 2.2 epa.compound Table,
Revised to Use Unique Compound id Column

Name	Formula	MW	cid	MP
Formaldehyde	CH_2O	30.03	1	−92
Guanidine hydrochloride	CH_6ClN_3	95.53	2	182.3
Dexamethasone	$C_{22}H_{29}FO_5$	392.47	3	262
Cortisone acetate	$C_{23}H_{30}O_6$	402.49	4	222
Phenobarbital	$C_{12}H_{12}N_2O_3$	232.24	5	174
Oxyphenonium bromide	$C_{21}H_{34}BrNO_3$	428.41	6	191.5
Metharbital	$C_9H_{14}N_2O_3$	198.22	7	150.5
Mesantoin	$C_{12}H_{14}N_2O_2$	218.26	8	135
Meperidine	$C_{15}H_{22}ClNO_2$	283.80	9	187.5
Vitamin D2	$C_{28}H_{44}O$	396.66	10	116.5

Table 2.3 epa.logP Table
Using Unique Compound id
Related to epa.compound Table

Temp	cid	logP	Method
25	1	0.35	exp
40	1	0.73	exp
	1	0.55	theory
	1	−0.11	theory
25	5	1.47	exp
25	6	0.17	exp
25	7	1.15	exp
	7	1.2	theory

original table? It might be possible to use the name column, requiring adding a name column to the logP table as well. Typically, a new column is added to each table containing a unique and arbitrary integer to establish the connection or relationship between the tables. In this case, the column could be called simply cid for compound id. The resulting tables are like Table 2.2 and Table 2.3.

Using a separate table to store logP information allows new values to be added as new rows in the logP table rather than as new columns in the original table. If one compound becomes particularly interesting and multiple measurements are made, these are easily stored in the logP table. Even hundreds or thousands of measurements of logP for one compound are no problem.

This kind of relationship between two tables is called a one-to-many relationship because for any one compound in the epa.compound table, there could be many rows in the logP table, related by the cid column. Once this type of relationship is established between two tables, it easily accommodates other tables of data. For example, if several molecular weight values for each compound were to become necessary, a new table for molecular weight could be created that would use the same compound id to relate to the compound table. The column used to relate two tables is called the key column. In this example, it would be called a primary key in the epa.compound table and a foreign key in the epa.logP table. Keys are discussed further in a later section of this chapter.

One-to-many is the most common type of relationship in relational databases. One-to-one relationships as well as many-to-many relationships are also possible and useful.

2.5.2 One-to-One Relationships

One-to-many relationships are common, but one-to-one relationships are everywhere. For example, in epa.compound Table 2.2, each compound has one formula. It would be possible to create a separate table for formulae, in which case there would be a one-to-one relationship between the epa.compound table and the epa.formula table. There is no fundamental need to separate molecular formulae into a separate table, simply because there is a one-to-one relationship. The relationship is based on understanding the nature of the data, namely understanding that a compound can have only one molecular formula. Nevertheless, data such as molecular formulae is sometimes stored in a separate table, for convenience, for clarity, or because it was added at a later date after the table was constructed.

Before breaking the original table into two tables, each compound also had just one logP. This was simply because not enough logP data had been collected or because multiple logP values were not yet of interest. In considering the design of a database schema, it is important to understand when a one-to-one relationship is an inherent attribute of the data. Consider the other two columns of the epa.compound table, molecular weight and melting point. It is possible to have multiple melting points, say, at different pressures, for each compound. It is also possible to have different molecular weights for different isotopes of each compound. If your needs dictate it, you should create a new table for melting points and molecular weights. If your needs are met by storing only one melting point and molecular weight per compound, leave these as columns of the epa.compound table. The rule of thumb is as follows: If there is a one-to-many relationship between two types of data, store the data in separate tables.

2.5.3   Many-to-Many Relationships

In the tables discussed earlier, the experimental or theoretical values were clearly attributable to one structure or compound. In some cases, say, for molecular weight, there was a one-to-one relationship with the compound. In the case of logP shown above, there were many values of logP and the separate table of logP values exhibited a one-to-many relationship with compounds. Consider the case of compound vendors. There are many vendors that supply any one compound and each vendor supplies many compounds. This situation is referred to as a many-to-many relationship between compounds and vendors. A separate table is created to define the unique vendors and the corresponding vendor ids. Along with the unique compound table, a third table is defined that contains only compound ids and vendor ids. The set of rows in this third table with a particular compound id indicates which vendors supply that compound. And the set of rows with a particular vendor id indicates which compounds that vendor supplies.

Many-to-many relationships are much easier to understand when visualization tools are used. Entity relationship diagrams (ERD) help visualize one-to-one, one-to-many, and many-to-many relationships.

2.6   Entity Relationship Diagrams

The definition of the tables and relationships in a database schema are completely described using the SQL language. This is discussed in a later chapter of this book. Using ERD is an excellent way to create and communicate a database schema. These can then be used to write the SQL necessary to create the tables in the database. There are many software tools available to create ERD and most of these can automatically output the proper SQL necessary to create the schema's tables and relationships. Some software tools even allow reverse engineering of existing schemas to create ERD. It is a good idea to begin working with ERD as early in the design process as possible.

The EPA schema described above in epa.compound and epa.logP can be represented as the ERD shown in Figure 2.1. In this figure, two tables are represented: epa.compound and epa.logP. The column names and data types for that column are listed. The PK symbol next to the cid column in epa.compound denotes that it is a primary key column. The FK symbol on the cid column of epa.logP denotes that it is a foreign key. The line joining the two tables shows the one-to-many relationship between the cid columns. The "crow's feet" symbol on the right side of the line near the epa.logP table shows that there are (possibly) many entries in the logP table for each compound, that is, for each unique cid in the epa.compound table. The circle, or zero symbol near the crow's feet indicates that there may be zero entries in the logP table for some compounds. The asterisk

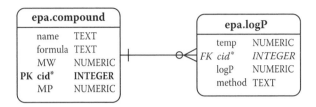

Figure 2.1 Entity relationship diagram showing the one-to-many relationship between compounds and logP.

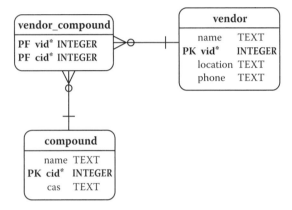

Figure 2.2 Entity relationship diagram showing the many-to-many relationship among compounds and vendors.

symbol after the cid data type means that these values cannot be null, since they act as primary and foreign keys. Other ERD diagram software may indicate keys and not-null values using other symbols, although the crow's feet symbol is a standard way of representing relationships between tables in most ERD software.

Many-to-many relationships are regularly encountered in chemical databases. In the case of many compound vendors and many compounds, the ERD in Figure 2.2 shows how three tables can be used to define a many-to-many relationship. The many-to-many type of relationship will be seen in other examples in later chapters. In Figure 2.2, the vendor table contains only as many rows as there are vendors. The compound table contains only as many rows as there are compounds. Each row in these tables contains the information available for each compound or vendor. No information is repeated in other rows or tables. The vendor_compound table contains many more rows, one row for each compound offered by each vendor. This table is the largest in this schema, yet it contains only integers, which are easily indexed and efficiently stored.

2.7   Uniqueness

In the table epa.compound, the name column is unique, meaning that no two rows have the same name. This might be coincidentally true for any column of any table, but in some cases the nature of the data requires that one column be defined as unique. Before declaring that a column in a table be unique, it is essential to understand the nature of the data. For example, the molecular weight is not a unique value as many structures share the same molecular weight, although the small set of data in table epa.compound happens to have only unique values of molecular weight. Molecular formula is also not a unique property of a molecular structure. It might be argued that name is not unique and indeed that are much better ways to uniquely identify molecular structure. However, since the purpose of Table 2.2 is to provide a primary table to store each molecular structure, it is advisable to have one unique column to prevent duplication of rows and possible confusion if the same structure is entered multiple times.

By defining the cid column in the epa.compound table, we artificially create a column that is unique. This cid is unique in the epa.compound table, but not unique in the epa.logP table. This is simply because the nature of the data in the EPA schema requires that each compound be "registered" in the epa.compound table, but it may have many logP values associated with it. Of course, many other tables analogous to epa.logP, for example, epa.solubility or epa.toxicity, could be added as those data become available or important in the database. The use of multiple tables, at least one of which defines the set of unique compounds of interest, is a hallmark of chemical relational databases. The use of multiple independent tables is one of the major advantages of RDBMS, allowing for easy extensibility.

2.8   Sequences

The cid column of table epa.compound is a simple integer, starting with 1 and increasing up to the number of compounds. The purpose of this column is to provide a unique key allowing tables in the schema to be related to one another. Any unique value would suffice, but integers are typically used because computers can store integers compactly and manipulate them efficiently. Any method of creating unique integers to be stored in the cid column would work, but most RDMS proved a convenient way to generate unique integers. The sequence function can be used to generate integers starting with 1 (or another chosen value) and increasing by 1 (or another chosen nonzero value). Every time a value is chosen from the sequence, that value becomes unavailable, ensuring a set of unique integers. There can be any number of sequences in the RDBMS. One typically defines a sequence that is associated with a unique column and not used for other purposes.

2.9   Keys

The importance of using one column of a table as a key was introduced in tables epa.compound and epa.logP discussed above. In these tables, the column named cid was defined as a key column. The term *key* is used to denote the column by which two or more tables are related to one another. This key column allows otherwise complex tables to be split into two or more tables.

2.9.1   Primary Keys

A primary key column in a table is also a unique column. There is often one central table in a schema by which the other tables in a schema are related to each other. In the EPA schema discussed here, epa.compound is the central table and the cid column is the unique, primary key. The purpose of the epa.compound table is to provide a central registry of compounds that are of interest in the schema. The key column is typically defined using a sequence to ensure uniqueness.

2.9.2   Foreign Keys

In the table epa.logP, the column named cid is also used. Its purpose is to relate the logP values to the compounds in the epa.compound table to which they refer. The cid column in epa.logP is clearly not unique since the table may contain multiple values of logP for any one compound and therefore uses the same cid value multiple times. Other tables analogous to epa.logP would use a column named cid in an identical way. A key column such as this, which is not unique and which relates to a primary key column in another table, is called a foreign key. The name of the foreign key column need not be the same as the name of its related primary key. The foreign key column is not defined by the sequence used to define the primary key, although it necessarily shares values with the primary key. The primary key column might be considered as "responsible" for the unique values relating the tables to each other. The foreign key columns are dependent upon the primary key.

There are many ways in which primary and foreign key columns can be used. A foreign key might also be unique, for example, in a table epa.casid that stores the Chemical Abstracts identifier. A table might contain two foreign keys, each of which relates to a primary key in different tables. A table might also contain both primary and foreign keys, each relating to keys in different tables. Proper use of entity-relationship diagrams becomes increasingly important as the complexity of the database schema and the table relationships grows.

2.10   Constraints

Uniqueness as discussed above is one kind of constraint that may be imposed upon the data values in a column. There are other kinds of constraints that are also useful. In a sense, just defining a column as an integer type constrains the values in that column to be integers. It may also be desirable to further constrain values to be positive. For example, in a column of molecular weight, such a constraint helps guarantee that nonsense values cannot be entered. It is even possible to define a constraint such that the molecular formula column could be used to compute a molecular weight consistent with the value stored in the molecular weight column. Judicious use of column constraints is one of the ways of ensuring the integrity of the data in the database. The exact nature of constraints and how they are specified in SQL is discussed in a later chapter.

2.11   Indexes

A column index can be created in order to speed access to the data in the column. For example, molecular weight might be an important column used in most searches of a mass spectroscopy database. If every row of the table has to be examined to determine if the value is between, say, 100 and 200, this would be slower than searching only those rows known to contain these values. One technique used to index columns of numerical data is to presort the values into a relatively small number of bins. The index records which rows belong to which bin. When searching for values between 100 and 200, only rows that belong to the appropriate bin need to be examined.

It is not necessary to do any sorting or other operations on the data to create the index. Every RDBMS has built-in indexing capabilities that apply to numerical, text, or even other data types. Most implementations of numerical indexing do not necessarily rearrange the table rows in a sorted order. The exact method used to create the index is not usually of interest to the database designer.

One very important use of the index is to speed up access to key columns. It is recommended to create an index on the primary and foreign key columns. This is important because it speeds up the methods used to relate the rows between the tables. This method is called joining the tables.

2.12   Joining Tables

When data is selected from a table, the purpose is to provide a subset of the table that is of interest. For example, rows may be selected where the molecular weight is below 500 and/or the logP is below 5. The result is itself a table with the chosen rows and columns from the table being searched. When data is stored in separate tables, the tables are conceptually (or perhaps actually, depending on the RDBMS) joined together into a larger table using the key columns to ensure the proper rows are combined with each other. In the EPA examples above, epa.logP rows would be joined with epa.compound rows where the columns cid have identical values. The result would be a table, for example, containing the name, molecular weight, and logP for selected compounds. For compounds with multiple logP values, there would be multiple rows each with a different logP value, but with the same name and molecular weight (and same compound id).

Efficiently joining tables requires careful consideration of primary key and foreign key columns, uniqueness, and indexing. While it is not necessary to use a primary key and its related foreign key column when joining tables, that is a very common, useful, and efficient way to join tables.

2.13   Normal Forms

There are many existing rules in SQL that prevent problems that cannot be prevented when using flat files. For example, a column of integers can only contain integers. In database theory, there are additional rules or suggestions designed to ensure that data tables operate properly and efficiently. One set of rules is referred to as normal form or normalization. These rules are not enforced by SQL, but it is a good idea to use these rules. This section will consider only the first three normal forms. Each form is more involved and more restrictive. There are at least six normal forms, but it is rare to encounter normalizations higher than three.

2.13.1   First Normal Form

A table is said to be in first normal form if each row has the same number of columns, each column has a value, and there are no duplicate rows. Because an RDBMS uses a table defined with a fixed number of columns, it is always true that each row contains the same number of columns. If one allows that null is a value, then every column will have a value. It should be obvious that repeating a row in a table is wasteful, but also potentially confusing and prone to error. For example, if two rows in a table of logP contained the same name and logP, one row may have the logP changed at some point. Then which row would be the correct row? This condition also illustrates the final aspect of first normal form: There should be at least one column, or combination of columns, that could function as a key that uniquely identifies the row. This is the name column or compound id column in the above examples. The data in this column must be unique.

It is not required that such a column actually be used as an SQL key, but it is wise to do so. In this way, the SQL uniqueness constraints can help to ensure that the table is in first normal form. While this one column must be unique, it is entirely possible and even likely that some other data values will not be unique. For example, there are expected to be many compounds that coincidentally have the same logP. This does not violate first normal form.

Another way to think about first normal form is to simply consider what the table is intended to contain. If it contains data about structures, then there must be a unique way to identify those structures, for example, by name. For a table that contains information about structures, it must contain only information about structures. It must contain simple values that are associated with that structure, for example, molecular weight or logP. It must not contain complex information about its data values, such as the method used to determine logP.

In the above example, there were cases where multiple logP values for a single compound became important. If these were inserted into the table of structures, this would violate first normal form. So they were removed to a second table related to the first. When there are multiple values, there should be a separate table that contains information about those values. The table of structures must not be used to contain information about its data values but only information about structures.

2.13.2   Second Normal Form

A table in second normal form must have data values that depend only on the key column that was identified while making sure the table satisfies first normal form. Using the EPA Table 2.1 example, suppose additional logP values needed to be recorded because measurements were made at various temperatures. If a logP_temp column were added, this would accommodate multiple logP values and the pair of columns, logP and logP_temp, taken together would be unique. But each logP value would depend on the logP_temp column, not just on the key column. This violates second normal form. The solution shown above in which logP values were removed to a separate table is the correct way to normalize to both first and second normal form.

One might be tempted to encode information into one column. For example, the logP values could be forced to fit into one column by encoding temperature along with the logP value, for example "1.55(25C)." While this would formally satisfy second normal form, this is a really bad idea. It would force the logP column to have to be a text column, thus eliminating the ability of the RDBMS to ensure numeric values of logP. It would further force users and database developers to follow new syntax rules and write functions to parse fields from within a column.

2.13.3 Third Normal Form

In third normal form, data values in a column are not intentionally repeated. For example, the separate logP table was created to satisfy first and second normal form. But it violates third normal form because the method values "exp" and "theory" are used repeatedly throughout the table.

The correct way to conform to third normal form is to create yet another table, here called method. Table 2.4 shows such a table. Table 2.5 shows how the logP table would be modified to use the method id instead of the text strings "exp" or "theory." Notice that this can solve one remaining problem with both Tables 2.3 and 2.5: there are two rows for compound 1 and method 2. This can arise when a new theoretical method becomes of interest in a project. Rather than inventing a new string, for example "theory2," it is a simple matter to add a new method to the method table in Table 2.4. Here it would be method 3. The logP table would be altered so that the appropriate rows for compound 1 would use method 2 and method 3. Furthermore, the method table can be made much more informative, perhaps explaining something about the experimental or theoretical methods, rather than just using cryptic handles such as "exp" and "theory" in Table 2.3. A more complete logP method table is shown in Table 2.6.

Some authors advocate analyzing schemas of tables up to normal form 3 and then backing off to second normal form in order to increase efficiency. With increasing capabilities of computers and RDBMS, this

Table 2.4 logP Method Table Containing Text Descriptions

Id	Description
1	Experimental
2	Theoretical

Table 2.5 logP Table Using Method id in Place of Text Methods

Temp	cid	logP	Method_id
25	1	0.35	1
40	1	0.73	1
	1	0.55	2
	1	−0.11	2
25	5	1.47	1
25	6	0.17	1
25	7	1.15	1
	7	1.2	2

Table 2.6 More Complete
logP Method Table

Id	Description
1	Hansch, C. et al. (1995)
2	BioByte clogP
3	gNova glogP v2.3

may not be necessary. The advantage of creating a method id and using a method table gives great flexibility and allows accurate and detailed description of each method.

2.13.4 Summary of Normal Forms

The following are not formal definitions of normal form, but it is hoped they will serve as reminders of the important reasons that these normal forms were originally suggested.

First normal form: Each table should contain only data about a unique entity. Each row should have a unique identifier. If data tables violate first normal form, a careful reconsideration of which information belongs in which table should be undertaken. A structure table should be about structures, a logP table about logP values, and a method table about methods.

Second normal form: Each row should contain only one value for each column. If multiple values of a data item are needed, a related table should be created for those values. Do not encode data "fields" into a data column. Create separate columns for each "field" and in a separate table, if necessary.

Third normal form: Do not repeat data values needlessly. Be wary of using codes, such as "exp" for experimental data values. But do not be afraid to violate third normal form at first and correct it when necessary. The Appendix shows a possible method for correcting violations of third normal form that might be encountered when importing data from another source, or for correcting violations that have crept into data tables over the lifetime of the database.

References

1. Codd, E.F. 1970. A relational model of data for large shared data banks. *Communications of the ACM*, 13(6):377–387.
2. U.S. Environmental Protection Agency. 2007. Estimation Program Interface (EPI) suite. http://www.epa.gov/oppt/exposure/pubs/episuite.htm (accessed April 21, 2008).

chapter 3

Structured Query Language (SQL)

3.1 Introduction

In Chapter 2, the concept of relational tables was introduced. In this chapter, the most common way of working with tables in an RDBMS is introduced. The SQL language provides ways to create tables, insert data, select data, delete data, update data, join tables, create table schemas, define functions, etc. SQL has many other features, not all of which are covered here.

3.2 Databases, Schemas, Tables, Rows, and Columns

The word *database* is used informally to refer to any collection of data. For example, one might call a file of e-mail addresses a database. There is a more formal definition of a database in an RDBMS. Information in an RDBMS is structured in a sort of hierarchy. An RDBMS contains databases, which contain schemas, which contain tables, which contain rows, which contain columns. A schema is a collection of tables. Rather than have tens or hundreds of tables in your database, the tables can be organized into schemas for clarity and convenience. In one sense, a schema is just a type of name space that allows a richer naming convention. For example, a database might contain several tables of structures. It might be convenient to use the table name structure to contain these structures. Rather than try to fit all the various structures together into a single table, or use coded table names such as nci_structures, pubchem_structures, or vendor_structures, separate schemas could be created. The fully qualified table names would then be nci.structure, pubchem.structure, and vendor. structure. The tables are segregated from each other by virtue of belonging to a separate schema, yet they are shown to be similar by sharing the same table name structure. Tables in different schemas are not isolated from each other; for example, they can be joined as readily as tables that belong to the same schema.

Schemas can also be used to contain functions, which are discussed later. For example, if there were several functions to compute the logP of a structure, it might be convenient to segregate the functions into separate schemas with fully qualified function names such as xlogp.logp, clogp. logp, or gnova.logp. In many ways, schemas function like folders or directories of files. In PostgreSQL, the default schema is called public. Any table or function created without specifying a particular schema belongs to the public schema. In Oracle, schema names have traditionally been associated with user names, but this is not part of the SQL standard nor is it required by Oracle.

A database can be thought of as a collection of schemas. It is possible to have many databases managed by one RDBMS, but each database is independent of any other. SQL was not designed to facilitate access to data in different databases. Recently, methods such as dbSwitch[1] or dblink[2] have made it possible to link together different databases. However, these are not considered here because they do not conform to the SQL standard and are implemented is various ways in different RDBMS. In the examples in this book, all schemas, table, functions, etc., are contained within one database.

3.3   Create

To create a schema named achemcompany, use the following SQL command:

```
Create Schema achemcompany;
```

To create a table using SQL, the name of the table is required along with the names and data types of the columns making up the table. Consider the following SQL command:

```
Create Table achemcompany.structure (
 smiles Text,
 id Integer,
 mw Numeric(6,2),
 added Timestamp(0));
```

This creates a table of four columns in the schema achemcompany. The column named smiles is intended to store the SMILES representation of a chemical structure, the id column will store an integer identifier to be used for joining other tables, the column mw will store the molecular weight with a precision of 2 digits to the right of the decimal point, and the column named added will record when this structure was entered into the table. As defined above, any character string could be entered into the smiles column, any integer into the id column, and any valid

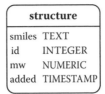

Figure 3.1 Simple entity relationship diagram for structure table.

timestamp into the added column. There are ways to improve the integrity of the data in this table, using uniqueness constraints, sequences, and user-defined functions to ensure proper SMILES syntax. These are discussed in later chapters. A simple entity-relationship diagram for this table is shown in Figure 3.1.

SQL is case-insensitive. Table names structure, STRUCTURE, or Structure each refer to the same table. Likewise, the command Create is the same as the command CREATE or create. If you wish to use case-sensitive names for table, schema, or column names, use double quotes, for example:

```
Create Table "aChemCompany".structure (
  smiles Text,
  "ID" Integer,
  "MW" Numeric(6,2),
  added Timestamp(0));
```

The data in the table is stored exactly as entered, preserving upper and lower case. SQL command keywords, such as Create and Integer, are arbitrarily shown in this book with an initial capital letter when SQL commands are shown.

3.4 Insert

Data is inserted into a table using the SQL insert command. For example:

```
Insert Into achemcompany.structure (id, smiles, mw, added)
  Values (1001, 'CC(=O)OC', 74.09, current_timestamp);
```

Notice that single quotes are required for text values and that the built-in SQL function current _ timestamp can be used to supply a valid timestamp. The order of the columns need not be the same as the order used in the create command. However, the order of the data in the Values clause of the Insert command must correspond exactly with the order of the columns as named in the command. There is a short-cut syntax of the Insert command that does not require column names. It should be avoided in favor of the syntax example above. There are other

Table 3.1 A Simple Structure Table

Smiles	id	mw	Added
CC(=O)OC	1001	74.09	2007-06-20 13:35:32
NCC(=O)OC	1002	89.11	2007-06-20 13:38:05
C(N)CC(=O)OC	1003		2007-06-20 13:38:21

ways of getting large amounts of data into tables, using the SQL Copy command or bulk-loading programs. These are not discussed here, but examples of using the Copy command are shown in Chapter 11 and the Appendix.

A sample table made using the above SQL Create and Insert commands is shown in Table 3.1. Note that more rows have been added to this table, not just the single row added with the Insert command above.

3.5   Select

Once data is inserted into a table, chosen rows and columns can be selected. For example, the following SQL command:

```
Select smiles,mw From achemcompany.structure Where mw < 100;
```

selects smiles and molecular weight from all rows that have molecular weight below 100. The Where clause of the SQL command can be quite complex, involving many comparisons of many columns from many tables.

An essential use of the Select command is to select data from different tables using the joining capabilities of RDBMS. Suppose there is another table of assay data defined using

```
Create Table hiv1.prot_inh (
 id Integer,
 ic50 Float,
 ki Float,
 tested Timestamp(0));
```

When data is inserted into this table, the proper id from the structure table is associated with the experimental data. In order to select structures with chosen inhibition constant K_i the hiv1.prot_inh table is joined with the achemcompany.structure table. For example:

```
Select smiles,ki
 From achemcompany.structure Join hiv1.prot_inh
 On hiv1.prot_inh.id = achemcompany.structure.id
 Where ki < 0.5;
```

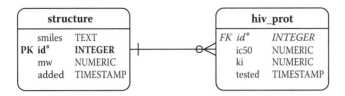

Figure 3.2 Entity relationship diagram for structure and hiv_prot tables.

Both tables are named in the SQL command, the chosen K_i limit is given and the `Join` condition `On hiv1.prot _ inh.id = achemcompany. structure.id` ensures that the proper rows of each table are joined. Note that the `hiv1.prot _ inh` table resides in a schema different from the `achemcompany.structure` table. Of course, this is not a requirement but is done for convenience and clarity during the design of the database. It is possible to leave out the schema name in the `On` clause because there is no ambiguity of table names. It is also possible to include the schema and table name along with the columns named `smiles` and `ki`. It is simply a matter of programming style. However, the schema name must be used for the table names in the `From` clause. Figure 3.2 illustrates the two tables and the relationship between them using the column named id.

There is an alternative form of SQL that is commonly used to `Join` tables. This alternative form is so common, especially in older SQL, that it must be mentioned here. The following SQL accomplishes exactly the same join described above, but uses a different syntax.

```
Select smiles,ki From achemcompany.structure, hiv1.prot_inh
  Where hiv1.prot_inh.id = achemcompany.structure.id And ki < 0.5;
```

Notice that the `Join` keyword is not used, but rather just a comma separates the two tables being joined. The `Join` condition becomes a part of the where clause. The former syntax using the `Join` keyword will be used in examples throughout this book.

3.6 Update and Delete

Once data has been inserted into a table, it may become necessary to update it. Suppose, for example, that a compound molecular weight had been entered incorrectly, or not at all for some particular compound. The following SQL command would update one row in the structure table.

```
Update achemcompany.structure Set mw=103.14 where id=1003;
```

Notice the use of the `Where` clause, similar to its use in the `Select` statement. It is important to use `Where` with `Update`. Otherwise every row of the table will be updated! This is almost never intended, but can be useful when necessary.

The following SQL command will delete one row from the structure table.

```
Delete from achemcompany.structure where id=1002;
```

Again, it is important to use a Where clause in the delete statement to specify precisely which rows are to be deleted and to prevent accidental deletion.

There are many other SQL commands besides Create, Insert, Update, Delete, and Select. For example, the Alter command modifies tables, columns, schemas, and other aspects of the database. There are many books that describe the SQL language. Most are specific to one particular RDBMS, such as PostgreSQL or Oracle. Many examples in this book will work in PostgreSQL, Oracle, MySQL and any other RDBMS that follows the SQL standard. However, some examples use data types and functions that differ among the various RDBMS. In those examples, PostgreSQL is used. In other words, all examples in this book will operate correctly using PostgreSQL. Most examples will work with Oracle and MySQL as well. Similarities and differences among various RDBMS are discussed in Chapter 4.

3.7 SQL Functions

There are many SQL functions available to perform common operations on data values. For example, sqrt, sin, and abs operate on numerical data. Text data can be operated on using functions such as substring, lower, and trim. These functions are explained in any book on SQL, although many of them are self-explanatory. This section will show how to create new functions for use in SQL. These functions may be written using SQL itself, or using the various procedural languages such as plsql, plpgsql, or plpython, depending on which RDBMS is being used.

There is another class of functions called aggregate functions. There are several standard SQL aggregate functions, such as sum, max, and avg. These are called aggregate functions because, rather than operating on a single value such as sqrt, they operate on multiple values. The result is an aggregate value such as a sum, maximum, or average of the multiple input values. This section shows how to use the standard SQL aggregate functions and explains how to create new aggregate functions for use with SQL.

3.7.1 Regular Functions

One type of SQL function is simply a collection of SQL statements. The input data type must be defined along with the data type of the result of the function. For example, the following SQL function converts pressure in atmospheres to kilopascals.

```
Create Function convert.atm_to_kpa(Numeric) Returns Numeric
  As 'Select $1 * 101.325;' Language SQL;
```

Once this function is created, it can be used just like any standard SQL function. This function uses PostgreSQL syntax.

As with tables, functions are associated with a schema within the database. In the above example, the schema named *convert* is used in order to conveniently locate conversion functions in a common schema. A function name can be fully qualified using its schema name, as in the following example.

```
Select bp, convert.atm_to_kpa(bp_press) as "kPa" from epa.properties;
```

The result is a table of boiling points and the corresponding pressure converted to kiloPascals from the pressure value stored in atmospheres in the table epa.properties.

Of course, more elaborate functions could be written, using SQL or one of the procedural languages available for the RDBMS being used. Plsql or plpgsql allow all of the SQL commands as well as structured programming constructs such as loops and if-then-else branching not available in SQL itself. When using languages other than SQL, keep in mind that the differences among various RDBMS is greater for the associated procedural languages than for SQL proper. The appendix of this book contains many examples of SQL functions written using SQL, plpgsql, plpython, and C.

3.7.2   Aggregate Functions

Aggregate functions operate on a group of values rather than individual values as ordinary (or scalar) functions do. SQL has several standard aggregate functions, for example, sum, average, and max. The following SQL would likely return multiple rows.

```
Select ic50 From hiv_inh;
```

The following SQL would return just one row.

```
Select avg(ic50) From hiv_inh;
```

Another use of an aggregate function depends on the use of the Group by clause of SQL. The following SQL will return multiple rows.

```
Select id, avg(ic50) From hiv_inh Group By id;
```

Each row will have the average ic50 for each compound id in the table.

It is possible to define a new aggregate function. One such function, Orsum, is shown in the Appendix. This function operates like the sum aggregate function, which produces the numeric sum of each member of the group (aggregate) specified in the SQL statement in which it is used. Orsum returns the logical OR of each bit string value, producing an aggregate bit string value representing bits that are set to 1 in any member of the group. This is used in the fingerprint and fragment key functions discussed later in this book.

3.8   Domains, Triggers, and Views

There are other features of SQL that are useful for chemical relational databases. Domains, triggers, and views are objects that belong to a schema just as tables and functions. These are also discussed in later chapters that focus on practical uses.

An SQL domain is an extension of one of the built-in data types, but includes an optional check constraint. For example:

```
Create Domain smiles As Text Check (valid(Value));
```

defines a Domain named smiles that is represented internally using the SQL Text data type. It must also satisfy the constraint that it is a valid SMILES. The valid(Value) function must return true or false. The argument Value passes the text string to the valid function. If the valid function returns false, an SQL error will be reported and the value will not be allowed. An example of the valid function using SMILES is given in a later chapter.

An SQL trigger is also useful to ensure that only valid data appears in a table. For example:

```
Create Trigger standardize Before Insert Or Update On atable
 For Each Row Execute Procedure standardize();
```

defines a Trigger named standardize, which causes a procedure named standardize to execute before any row is Inserted or Updated in the table named atable. This trigger may actually correct invalid data, if possible. Several examples of trigger functions are given in later chapters.

An SQL view is very similar to a table. It has rows and columns of defined data types just as a table. A view is defined by selecting particular rows and columns from one or more tables, using an SQL select statement. For example:

```
Create View test_set (logp, temp) As
 Select logp, temp From literature_data Where ref Like 'Hansch%1995%'
 And temp Is Not Null;
```

creates a view called `test _ set`, which is a subset of the table named `literature _ data`. The test_set will only contain rows with octanol-water partition coefficients reported by Hansch in 1985 and having a temperature reported. The `test _ set` view can be used as if it were a table in other `Select` statements. It is not possible to update or delete rows from a view. Several examples of views are given in later chapters.

3.9   Unions, Intersections, and Differences

The results of a `Select` statement are in the form of a table. This can be a subset of a single table, or the result of joining several tables. The exact set of rows is chosen by using various `Where` clauses. The use of Boolean operation such as `and`, `or`, and `not` allows a sort of union (or), intersection (and), and difference (not). For example:

```
Select logp From logp Where temp = 25 And
 ref Like '%Hansch%' Or ref Like '%Yalkowsky%';
```

produces a different set of rows than the following:

```
Select logp From logp Where temp = 25 And
 (ref Like '%Hansch%' Or ref Like '%Yalkowsky%');
```

The careful use of the Boolean operations `and`, `or`, and `not` along with parentheses will produce the desired set of rows from any table. When data are in separate tables that are related to one another, this approach also works well when the two tables are joined together using the SQL `Join` clause.

When data are selected from tables that are not related to one another, a different approach is used. The SQL operators `union`, `intersect`, and `exclude` allow set operations on tables or the sets of rows resulting from a select statement. For example:

```
Select logp,temp from logp Where ref Ilike '%Hansch%' And temp = 25
  Union
Select logp,temp From merck Where temp Is Not Null;
```

produces a set of rows from two unrelated tables, logp and merck. It is necessary that the number and data types of the columns from each select statement be identical. Any number of `Select` statements may be combined using this method. The `union`, `intersect`, and `except` operations can be mixed in any order, using parentheses as necessary to effect the correct overall Boolean operation. It is possible to use this method to combine results selected from the same table or from related tables. In those cases, it is possible to craft two different SQL statements—one using `intersect`, union and exclude operations and the other using and, or, and not where clauses—that yield the same result. One or other of the two statements may be preferred for speed of execution, clarity of expression, or ease of extensibility.

References

1. Dar, S., Hecht, G., and Shochat, E. 2004. Industrial sessions: Database applications: dbSwitch™: Towards a database utility. In Proceedings of the 2004 ACM SIGMOD International Conference on Management of Data SIGMOD, ed. Weikum, E., König, A.C., and Deßloch, S. Association for Computing Machinery, pp. 892–896.
2. PostgreSQL 8.2 documentation. 2008. Dblink. http://www.postgresql.org/docs/current/static/dblink.html (accessed April 21, 2008).

chapter 4

Relational Database Management Systems

4.1   Introduction

There are many relational database management systems (RDBMS) available. A full comparison of every feature of every RDBMS is beyond the scope of this book. Such comparisons quickly become outdated. A search using Google or Wikipedia is a good place to start if you want to make a comparison. A comparison of Oracle, MySQL, and PostgreSQL is available from the Computing Division at Fermilab.[1] As with any comparison, be sure to note which versions of each database are being compared. Comparisons are most valuable when the most recent version of each database is considered. Besides objective feature comparisons such as the above, other considerations are important. Sometimes a particular database is already being used at a particular company or research institution. This can be a great advantage, considering that support and advice of colleagues is often quick and highly relevant.

Much of this book discusses ways in which the RDBMS can be used and even extended to handle chemical structures correctly, quickly, and conveniently. Extensions of the capabilities of PostgreSQL are simply called *extensions*. Oracle uses the term *data cartridge*. There are chemical extensions or cartridges available for PostgreSQL, Oracle, and MySQL.

PostgreSQL is a free and open-source RDBMS that traces its roots to Ingres, one of the first RDBMS created. It strongly conforms to the ANSI-SQL-92/99 standards. There are several independent companies that offer support for PostgreSQL. There are commercial products that use PostgreSQL as an underlying database. CHORD, a commercial chemical cartridge for PostgreSQL, is sold by gNova.[2] PgChem, an open-source cartridge for PostgreSQL, is available at SourceForge.[3] RDKit also contains an open-source cartridge for PostgreSQL.[4]

Oracle is a commercial RDBMS. It strongly conforms to the ANSI SQL-92/99 standards. It is the most widely used RDBMS. The Oracle company sells and supports the Oracle RDBMS. There are many other companies that also offer support for Oracle RDBMS. There are several commercial chemical cartridges available for use with Oracle.[5]

MySQL is an open-source RDBMS. It conforms in part to the SQL 92/99 standards. The MySQL company sells and support the MySQL RDBMS. There are other companies that support MySQL and offer products that use MySQL as the underlying database. Tharun Kumar Allu describes a small molecule chemical database cartridge extension for MySQL.[6]

4.2   Standard SQL

SQL was first standardized by ANSI in 1986. Updated standards are referred to as SQL92, SQL99, and SQL2003. Most RDBMS conform to the SQL92 standard, including some features from later standards. Most also provide additional features not addressed in standard SQL at all. Sometimes the differences among the RDBMS are simply minor syntax differences, but sometimes there are more fundamental differences.

No RDBMS conforms exactly and completely to any SQL standard. For this reason, books on SQL almost always concentrate on one particular RDBMS. One notable exception documents every SQL command and details differences across different RDBMS.[7] While that book discusses differences in detail, this book discusses concepts of RDBMS in general, so that any RDBMS could be used to implement the methods described here. When specific SQL examples are given however, the SQL statement uses PostgreSQL syntax. Many times this is identical to the syntax for other RDBMS, or there is only a minor difference.

Some of the more advanced methods described in this book require a more specific use of the RDBMS. The choice made for this book is PostgreSQL. In cases where a particular feature of PostgreSQL is used, a note is added to alert the reader. For example, the array data type in SQL2003 is implemented in PostgreSQL very differently than in Oracle. The `list _ matches` function described in a later chapter of this book returns an array of integers that denote which atoms in a structure match a substructure query. The integration of this function into SQL would be handled quite differently in PostgreSQL, Oracle and MySQL.

4.3   A Sampling of Differences

It is not feasible to detail every difference among the various RDBMS. There are several commonly encountered differences that merit some attention here.

The text data type can be defined using the keyword text, according to the SQL99 standard. This is implemented in both PostgreSQL and Oracle, although one rarely sees this in Oracle. Instead, the `varchar` or `varchar2` keyword is used. In PostgreSQL, the `character varying` or `text` keyword is common, as well as `varchar`. At first glance, this difference may seem minor: changing one keyword to another is a simple thing to do except for a large and complex database application. But another difference requires more careful consideration.

In Oracle, the string data type requires a specification of the maximum length of the text, for example `varchar2(1000)`. This is possible, but unnecessary in PostgreSQL where the string data types can be of unlimited size. Depending on the point of view, it may seem liberating to be unconcerned about maximum text length, or it may seem dangerous to allow text of arbitrary length. Another consideration in using the text data type is how the null value is treated.

In standard SQL, a text column may contain data, or not. The `null` value is used to denote that no value is contained in that column of any particular row. The `null` value can be used in SQL statements, for example:

```
Select smiles from nci.structure where name is null;
```

In Oracle, an empty text string is considered equivalent to the `null` value. In PostgreSQL, an empty string is a valid, nonnull value. Neither approach is more correct than the other, but it may require some programming adjustments when moving from PostgreSQL to Oracle, or the other way around.

Aside from standard SQL, RDBMS typically include a procedural language that builds upon SQL. Oracle calls this plsql. PostgreSQL calls this plpgsql. These languages include loops and conditionals that are not part of the SQL standard language. Unfortunately, the differences among the procedural languages are much greater than the differences among the various RDBMS implementations of the SQL standard language. One advantage offered by PostgreSQL is a wider variety of procedural languages, namely plpython, plperl, and pltcl. In addition, PostgreSQL allows the SQL language to be extended using C language functions. These functions operate just like internal standard SQL functions. The standard internal function library is a rich source of coding examples for user-defined C functions.[8]

4.4 Server and Client

The RDBMS is installed and runs on a computer that functions as a database server. Any SQL commands are executed on the server by the RDBMS. Functions written in SQL or in any of the procedural languages mentioned above are also executed by the RDBMS. This has the advantage that the data tables used by these SQL commands or procedural functions are under the control of the server. This is the most efficient way to access the data. The disadvantage is that the server may have many requests to handle from many users. Another way to operate on data tables is indirectly, using a client program typically (although not necessarily) run from another computer.

A client program communicates with the RDBMS server using a TCP port. The details of that communication are not important for the purposes of this book. Various client program libraries handle those details. The advantage of using a client program is that any number of client computers can be used to spread the workload. The disadvantage is that data from the server must be delivered to and from the client computers, resulting in some inefficiency.

As with any client/server architecture, the efficiency concerns are not easy to accurately predict. Consider a simple example where one wishes to find all data values less than 1.0. A simple SQL command would accomplish this, but would run on the server. Only the resulting subset of data would be delivered to the client. Using an extreme approach, a client would require the entire data set and then select the values less than 1.0. Which approach is more efficient depends on many things: the size of the data set, the size of the subset less than 1.0, the relative efficiencies and loads of the server and client, and the speed and latency of the data channel linking the server and client. For another operation that is more complex than selecting values less that 1.0, the degree of complexity is an important consideration. Clearly, there is no simple "one size fits all" solution to deciding whether to use a server or client program.

The standard way of sharing data between a client and an RDBMS server is to use the SQL language. Many computer languages, such as perl, python, java, and C have standard libraries or methods that use SQL to read and write data to and from their internal data types and data structures. Many other computer programs, such as Excel, OpenOffice, and R have SQL "hooks" to allow transfer of data to and from an RDBMS server. This can be a great advantage when the computer client program already provides most of what is required. For example, if the goal of a project is to provide users with an Excel spreadsheet of data, why not use Excel directly? If a project already uses a number of numerical analysis programs written in R, why not use R's ability to interface with the RDBMS? If most of the programmers on a particular project are fluent in java, why not use jdbc (a standard javan RDBMS interface)? Later chapters of this book present specific examples of client programs that use SQL.

There is a smaller set of tools that are typically run on the server. Any SQL commands and any procedural language functions are run on the server. In principle, there is complete flexibility of the server side tools, since in principle any computer program can be written in any computer language. Later chapters of this book show how the RDBMS server itself can be extended using server side programming to handle chemical information. These extensions may directly solve the needs of a particular project, but more importantly they increase the flexibility of the RDBMS to handle chemical information. Client programs can use the results of chemical searches and other computations as well.

4.5   Compatibility

There are many client programs available that allow one to access data in an RDBMS. Some work properly only with one particular RDBMS, while others strive for compatibility various RDBMS. Sometimes there are different versions of the same program, each intended for use with a different RDBMS. Many computer languages have modules that allow access to data in an RDBMS. Most have modules that include an abstraction layer that allows programming without regard to the exact RDBMS being used. An underlying layer then implements the particulars of interfacing with each RDBMS. These issues are discussed more fully in Chapters 5 and 12, along with examples using various computer languages.

References

1. Fermilab Computing Division. 2005. Comparison of Oracle, MySQL and PostgreSQL DBMS. http://www-css.fnal.gov/dsg/external/freeware/mysql-vs-pgsql.html (accessed April 21, 2008).
2. O'Donnell, T.J. 2005. CHORD extension to PostgreSQL. http://www.gnova.com (accessed April 21, 2008).
3. *Schmid*, E. 2004. Pgchem::tigress extension to PostgreSQL. http://pgfoundry.org/projects/pgchem/ (accessed April 21, 2008).
4. Landrum, G. 2006. RDKit: Cheminformatics and Machine Learning Software. http://rdkit.sourceforge.net/ (accessed April 22, 2008).
5. Daycart. 2008. http://www.daylight.com (accessed April 21, 2008); Jchem cartridge. http://www.chemaxon.com (accessed April 21, 2008); CambridgeSoft Oracle cartridge. http://www.cambridgesoft.com (accessed April 21, 2008); MDL Isentris. http://www.symyx.com (accessed April 21, 2008)
6. Kumar, T. 2005. Small molecules database cartridge. http://www.unm.edu/~tharun/smdb.html (accessed April 21, 2008).
7. Kline, K. 2004. *SQL in a Nutshell, 2nd Edition*, Sebastopol, CA: O'Reilly.
8. PostgreSQL 8.2 documentation. 2008. C-language functions. http://www.postgresql.org/docs/8.2/static/xfunc-c.html (accessed April 21, 2008).

chapter 5

Client and Web Applications

5.1   Introduction

Once a database has been installed, there are many ways to interact with it. The primary way is using structured query language (SQL), either directly or indirectly. Some client programs connect to the database server directly and allow the user to type in SQL commands and display the results with minimal processing. Some programs are more elaborate, providing a Web-based interface or other GUI (graphical user interface). These applications typically provide some amount of postprocessing of SQL output. In some cases, many operations can be carried out without direct knowledge of SQL. These will certainly help the novice database user, but may also satisfy many of the needs of more experienced users and developers.

Sometimes these general-purpose applications are not sufficiently specific to the needs of database users. In that case, custom applications can be written. Many programming languages have extensions that allow data to be selected from the database and read into data structures for further operations. This chapter considers ways of using ODBC (Open Database Correctivity), JDBC for Java, Perl::DBI (Database Interface), pg and pgdb for Python, and PDO for PHP.

5.2   Command Line Programs

Command line programs are one simple way to interact with the database server. Each RDBMS provider supplies a command line program. For example, PostgreSQL supplies psql, Oracle supplies sqlplus, and MySQL supplies mysql. These command line shells allow the user to type in SQL commands. The results are displayed, usually with minimal formatting. There are other non-SQL commands that are unique to each RDBMS. For example, the PostgreSQL psql command \d mytable will describe the nature of the table named mytable, showing the names and data types of the columns. The corresponding Oracle plpsql command is describe mytable. For simple inquiries, command line SQL shells such as these are very useful.

Here is a sample SQL command and output from the psql command line client for PostgreSQL.

```
> select sid, activity_outcome, "log_gi50_M" from pubchem.nci_h23
  limit 10;
    sid | activity_outcome | log_gi50_M
  -------+------------------+------------
  67107 |                1 |         -4
  67121 |                2 |     -7.287
  67122 |                1 |     -4.688
  67213 |                1 |         -4
  67217 |                1 |         -4
  67294 |                1 |     -5.339
  67379 |                1 |         -4
  67383 |                1 |         -4
  67432 |                1 |         -4
  67525 |                1 |     -4.743
  (10 rows)
```

For routine database maintenance such as backup, each RDBMS provider typically supplies other command line programs. For example, PostgreSQL provides the pg_dump program that outputs a file of SQL commands containing the definition and data contained in each table, schema, etc., in the database. This file becomes a backup of the database, which can be restored using the psql command to execute the SQL commands in the file.

Depending on the level of familiarity with command shell programs, these can become the primary method of working with a database. Programs can be written in any language to produce SQL commands that are then passed to the SQL command shell. However, for browsing and formatting data, other methods will probably be more suitable.

5.3   Web-Based Applications

Because of the flexibility and familiarity of Web browsers, Web-based applications have become very popular and powerful. There are many Web-based interfaces to RDBMS. For example, phpPgAdmin[1] is a popular Web application for PostgreSQL. PhpMyAdmin[2] is popular with users of MySQL. These applications allow one to connect to a chosen database server, browse the schemas and tables in the database, and enter SQL commands. The output from SQL commands, especially table output, is formatted nicely. Other operations, such as creating, altering, or dropping tables and schemas, are also provided. These use an HTML form interface, with text boxes, radio buttons, check boxes, and other form elements familiar to all uses of the Web.

One advantage of Web-based applications is that the Web browser can be run on virtually any desktop or laptop computer. The database server itself can be located elsewhere, use a different operating system, and be maintained by others. In addition, the Web server can reside on yet a third

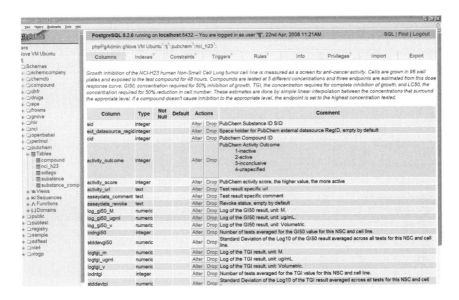

Figure 5.1 Typical Web page using phpPgAdmin.

computer. Of course, the Web browser, Web server, and database server might even be on a single laptop. The flexibility of this approach makes it easy to start simply and expand portions of the system as required.

Figure 5.1 shows a sample Web page from phpPgAdmin. The left frame is interactive, allowing the user to view and select various database, schemas, tables, functions, etc. The right frame typically shows table data, results of SQL commands, or interactive Web forms allowing operations on the database.

Figure 5.2 shows a typical RDBMS installation. The server is a Linux server using PostgreSQL, Oracle, or MySQL as the RDBMS, apache as the Web server, php as a Web application language, and psql, plsql or mysql as an SQL command shell. The client might be a Linux, Windows, or Mac laptop or desktop computer. It would use a Web browser, such as Firefox, Explorer, or Safari. A telnet client program would allow using the SQL command shell. Other applications, perhaps using ODBC or JDBC could also be used on the client. These types of applications are discussed in the next section of this chapter.

5.4 Client Applications

Other client applications are also available for use with an RDBMS. Compared to Web-based applications, client applications are more specific to the particular client desktop or laptop computer. For example, PgAdmin[3] is used to interact with PostgreSQL. Various versions of

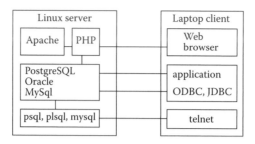

Figure 5.2 A typical client-server installation for an RDBMS.

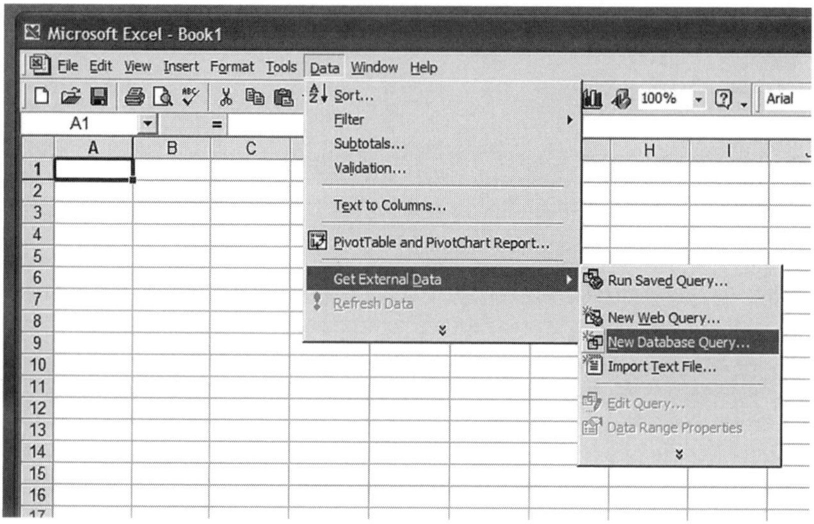

![Microsoft Excel screenshot showing Data menu with Get External Data submenu]

Figure 5.3 Excel Data/Get External Data/New Database Query.

PgAdmin are available to be run on Linux, Windows, Mac, and other clients. Oracle users typically use Oracle's SQL Developer or Toad (Toad also works with MySQL). SQL Developer[4] is a Java application that uses JDBC to mediate the communication to the database. A typical user need not be concerned with this, but a later chapter of this book discusses ways to construct new client applications using JDBC, ODBC, and other methods to communicate with the RDBMS. Toad is a Windows only application with a free version and a full-featured pay version.[5]

It is also possible to use Excel to select data from an RDBMS using ODBC. The ODBC connection is first set up using the server name, database name, and user login information. Then a dialog window is used to select the desired rows and columns from the database. The returned values are inserted into the Excel table. Figure 5.3 show a sample session using Excel to

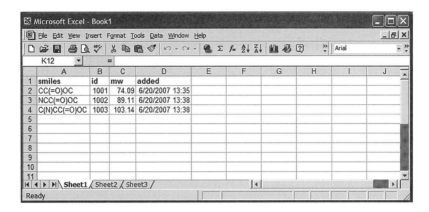

Figure 5.4 Excel spreadsheet with selected data.

process an SQL request for data from a table. This opens a dialog window by which the user selects columns and rows from tables in the database. It is also possible to have saved queries that users can select. In either case, the rows and columns of data are inserted into the Excel spreadsheet. Figure 5.4 shows the results in Excel after selecting the desired columns. Once this data is available in Excel, any further operations or formatting is possible.

Another useful client program is R.[6] It is used for statistical analysis of data and has some nice graphical capabilities as well. There is an add-on to R that uses ODBC to communicate with an RDBMS server.[7] Consider the following R program.

```
require("RODBC");
channel = odbcConnect("PostgreSQL30", uid="reader", pwd="something");
sql = "Select logp, xlogp From xlogp.test_set";
df = sqlQuery(channel, sql, max=0);
plot(df);
```

When this is run, the plot shown in Figure 5.5 is produced. With virtually no programming other than a simple SQL statement, a plot of two columns of data from a table can be produced using R. Of course, once this data is read into an R dataframe, many other complex statistical operations can also be performed. Some of these are discussed in Chapter 12.

Many other useful client programs allow input of data using SQL. For example Spotfire[8] and Pipeline Pilot[9] allow data to be read for an RDBMS using ODBC.

5.5 SQL Interfaces in Various Languages

When developing a client application, one or more computer languages will be chosen. The purpose of this section is not to advocate one language

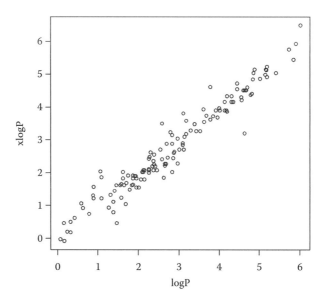

Figure 5.5 Plot using R of computed versus experimental logP values.

over another, but to show examples of how an SQL interface is accomplished in several common computer languages. In each of these examples, it is assumed that there is an existing PostgreSQL database available on a server named *rigel*. This database contains a schema named *nci* that contains a table named *structure*. The table has at least the columns *smiles* and *cas*, which are used in these examples. The database in these examples is named *book* and the user is named *reader*. In each of the examples, a database connection is established. This requires the name of the host database server, the database name, and the user name and password. The exact syntax of the statement creating the connection is different in each computer language.

In each example, a simple SQL statement is used to select rows that match a particular substructure. The rows are then simply fetched and printed. These examples should serve as a starting point to understand how client programs communicate with an RDBMS server using SQL. The examples can also server as a basis for other more complex client programs.

Selecting rows from a table is only one common operation needed for a client program to work with an RDBMS. In addition to a simple SQL select statement, other more complex select statements will become necessary as more complex client programs are developed. It is also important to see how to properly use the SQL Insert statement. Rather than showing examples in several computer languages, the use of more complex SQL statements is discussed in Chapter 12.

5.5.1   Perl

Perl is a general-purpose computer language with no built-in capability to interface with an RDBMS. However, there are readily available modules that enable Perl to use all of the common RDBMS. Two modules are needed. One is a general database interface called DBI. This allows one to write applications without having to deal with details specific to any particular RDBMS. The other module, DBD is specific to the particular RDBMS being used. There are several different DBD modules, for example DBD::Pg for PostgreSQL and DBD::Oracle. A simple way to install these is using CPAN, the comprehensive perl archive network. On a linux system, the following commands will install the two modules necessary to use perl to access a PostgreSQL database. These should be run as the linux super user by using the sudo command, or logging in as super user.

```
sudo cpan -i DBI
sudo cpan -i DBD:Pg
```

Once these modules are installed, the following perl script will select some rows and columns from a PostgreSQL database.

```
use DBI;
use DBD::Pg;
#these variable must be set to values appropriate for your site
my $dbname = "book";
my $username = "reader";
my $password = "something";
my $host = "rigel";

my $dbh = DBI->connect("dbi:Pg:dbname=$dbname; host=$host",
  $username, $password);
my $sql = "Select smiles,cas from nci.structure where
  matches(smiles,'c1ccccc1C(=O)NC')";
my $sth = $dbh->prepare($sql);
my $rv = $sth->execute;
while (my @row = $sth->fetchrow_array()) {
 print join "\t",@row;
 print "\n";
}
$dhb->disconnect;
```

The variable $dbh is referred to as a database handle. It is created by connecting to the database using the DBI->connect function. This statement is typically executed only once, although it is possible to connect to multiple databases or to connect and disconnect as needed. The variable $sth is referred to as a statement handle. It is typically used several times, once per SQL statement. The statement is prepared and then executed. The results are fetched and processed as required. There are many DBI functions to fetch results from an SQL statement. These are not considered here further, but are described elsewhere.[10,11]

5.5.2   Python

Python is a general-purpose computer language with no built-in capabilities to access an RDBMS. There are modules that extend python to allow interaction with an RDBMS, for example, pygresql[12] for PostgreSQL or cx_oracle for Oracle. Once pygresql is installed, the following python script will fetch some rows and columns from a PostgreSQL database.

```
import pg
conn = pg.connect(dbname='book', host='rigel', user='reader')
sql = "Select smiles,cas from nci.structure where
  matches(smiles,'c1ccccc1 C(=O)NC')"
for (smi,cas) in (conn.query(sql).getresult()):
 print smi, cas
conn.close()
```

The connect function opens the connection to the RDBMS using the appropriate database name, host name, username, and password. The query and getresult methods execute the SQL statement and get the results. There are other methods available to get results, but these are discussed elsewhere.[13]

5.5.3   PHP

PHP is a general-purpose computer language often used as a CGI program serving as part of a Web application. The interface to an RDBMS is compiled into PHP. When PHP is installed, the appropriate version must be selected to allow interaction with the necessary RDBMS. Of course, it is possible to compile PHP from source, specifying the options to include RDBMS support. PHP supports interaction with many RDBMS, including Oracle[14] and PostgreSQL.[15] Once the proper version of PHP is installed, the following php script will fetch some rows and columns from a PostgreSQL database.

```
 <?php
$dbconn = pg_connect("host=rigel dbname=book
  user=reader password=something");
$sql = "Select smiles,cas from nci.structure where
  matches(smiles,'c1ccccc1C(=O)NC')";
$result = pg_query($dbconn, $sql);
while ( $row = pg_fetch_array($result) ) {
  print $row['smiles'] . "\t" . $row['cas'] . "\n";
}
pg_close($dbconn);
?>
```

There are other methods to execute SQL queries and fetch results, but these are described elsewhere.[16]

5.5.4   *Java*

Java is a general-purpose computer language that includes the java.sql package.[17] This package is part of the standard java distribution from Sun and provides a generic interface to an RDBMS. JDBC (Java Database Connectivity) is a separate driver that enables communication between java.sql and a variety of RDBMS. The driver is in the form of a jar file. A separate driver is needed to enable java communication with each different RDBMS. The PostgreSQL JDBC driver[18] is used in the following example. It must be downloaded and installed in order for the following example to work properly.

```
import java.sql.*;

public class JDBCDemo {
  public static void main( String[] args ) {
    try {
      // Connect to the database
      Class.forName("org.postgresql.Driver");
      String url = "jdbc:postgresql://rigel/book";
      Connection con = DriverManager.getConnection(url, "reader");
      // Execute the SQL statement
      Statement stmt = con.createStatement();
      ResultSet resultSet = stmt.executeQuery("SELECT
        smiles,cas from nci.structure where gnova.
        matches(smiles,'c1ccccc1C(=O)NC')");
      System.out.println("Got results!");
      // Loop thru all the rows
      while( resultSet.next() ) {
        String smi = resultSet.getString( "smiles" );
        String cas = resultSet.getString( "cas" );
        System.out.println( smi + "\t" + cas );
      }
      stmt.close();
    }
    catch( Exception e ) {
      System.out.println(e.getMessage());
      e.printStackTrace();
    }
  }
}
```

Once this is compiled, it can be run as follows, being sure that the PostgreSQL JDBC jar file is in the class path. The exact location of the jar file may be different than in the example here.

```
java -cp .:/usr/share/java/postgresql.jar JDBCDemo
```

There are other methods available to prepare SQL statements and to fetch results. These are described in the standard java.sql documentation, in on-line tutorials,[19] and in various books.[20,21]

References

1. PhpPgAdmin. 2008. http://www.phppgadmin.com (accessed April 18, 2008).
2. PhpMyAdmin. 2008. http://www.phpmyadmin.net/ (accessed April 18, 2008).
3. PgAdminIII. 2008. http://www.pgadmin.org (accessed April 18, 2008).
4. Oracle SQL Developer. 2008. http://www.oracle.com/technology/products/database/sql_developer/index.html
5. Toad DBA Suite for Oracle. http://www.quest.com/toad-dba-suite-for-oracle/ (accessed April 18, 2008).
6. The R Project for statistical computing. 2008. http://www.r-project.org/ (accessed April 18, 2008).
7. RODBC database interface module. 2008. http://cran.r-project.org/web/packages/RODBC/index.html (accessed April 18, 2008).
8. Spotfire. 2008. http://spotfire.tibco.com/ (accessed April 18, 2008).
9. SciTegic platform. 2008. http://www.accelrys.com/products/scitegic/ (accessed April 18, 2008).
10. Perl DBI module. http://search.cpan.org/~timb/DBI/DBI.pm (accessed April 18, 2008).
11. Matthew, N., and Stones, R. 2005. *Beginning databases with PostgreSQL: From novice to professional. 2nd ed*. New York: Apress.
12. PyGreSQL – PostgreSQL module for Python. 2006. http://www.pygresql.org/ (accessed April 18, 2008).
13. PyGreSQL programming information. http://www.pygresql.org/pg.html (accessed April 18, 2008).
14. Oracle OCI8 functions. 2008. http://us2.php.net/oci (accessed April 18, 2008).
15. PostgreSQL functions. 2008. http://us3.php.net/pgsql (accessed April 18, 2008).
16. Pg_fetch_result function. 2008. http://us.php.net/manual/function.pg-fetch-result.php (accessed April 18, 2008).
17. Package java.sql. 2003. http://java.sun.com/j2se/1.4.2/docs/api/java/sql/package-summary.html (accessed April 18, 2008).
18. PostgreSQL JDBC driver. 2008. http://jdbc.postgresql.org/ (accessed April 18, 2008).
19. Creating complete JDBC applications. 2008. http://java.sun.com/docs/books/tutorial/jdbc/basics/complete.html (accessed April 18, 2008).
20. Bales, D.K. 2001. *Java programming with Oracle JDBC*. Sebastopol, CA: O'Reilly.
21. Melton, J. and Eisenberg, A. 2000. *Understanding SQL and Java together: A guide to SQLJ, JDBC, and related technologies*, San Francisco: Morgan Kaufmann.

chapter 6

Data Storage, Searching, and Manipulation

6.1 Introduction

A schema is a collection of tables and functions in a database. There is no single schema that will satisfy the needs of every chemical database user. It might be possible to use an existing schema, perhaps one from this book, and modify it to suit the needs of a particular project. It might be necessary to examine the needs of the project and develop an entirely new schema. The purpose of this chapter is to give examples of useful schemas and to provide enough background to allow the design of new schemas.

In Chapter 2, the usefulness of relational tables was introduced. Sample data from the U.S. Environmental Protection Agency was used to show the advantage of storing each type of data in a separate table. The data in each table remain related to the proper chemical compound through the use of a unique chemical id, which functions as a unique key relating multiple tables. This technique will be used extensively in this and following chapters. The separation of data into multiple tables also facilitates cases where a compound may have multiple data values, also known as one-to-many relationships. This chapter will show examples of how many-to-many relationships are handled. It will also show more examples of how the choice of data types affects the operation of the database and the applications that use it.

6.2 General Schema Design Decisions

When designing a schema to hold chemical information, it is crucial to first consider how the data will be used. One approach is to interview potential users of the database to determine what questions need to be answered on a regular basis. For example, users of a chemical compound tracking system will typically need to know the following:

- Where is compound X now?
- Which compounds does chemist Z have checked out?
- Has compound Q already been registered?
- How many samples of compound Y have been prepared?

It is essential that the schema of tables be created in a way that can easily answer such questions. During the design of a schema, it is useful to prepare structured query language (SQL) queries that will provide the answers to these questions. If it proves difficult or awkward to use the schema to answer these questions, the schema must be redesigned. Do not underestimate the importance of time and effort spent during the design of a database schema.

Another task in selecting or designing a schema is to determine which operations need to be performed. For example, in a compound-tracking system, compounds need to be registered and samples of compounds need to be taken (checked out). For a schema to function well, it should be relatively simple to update or insert data into the tables of the schema to record these activities.

It is worthwhile to also consider how important each question or operation is, or perhaps how frequently it will be required. Be sure to determine which operations are essential to the smooth functioning of the system and which questions need to be answered quickly. In other words, prioritize the requirements since it will rarely be possible to satisfy all the requirements without some compromise. Finally, consider questions or operations that might not be needed immediately but will possibly be required in future.

Once these questions and operations are stated, one needs to consider the items of information that must be stored in the tables of the schema. These will typically be words or concepts used in the set of questions posed during the design of the schema. For example, in a chemical compound-tracking database, some items would be samples, compounds, chemists, locations, and checkout and registration events. Sometimes these items are simple, such as molecular weight or a chemist's name. These can be represented using simple SQL data types, such as numeric or text. Often, these items are complex, such as location or sample. These items are represented by defining a set of columns (a table) that contains simple SQL data types, or possibly even references to other complex data types. The compound-tracking example is considered in detail in a later section of this chapter.

Of course, it will be important to store the chemical structure itself. Clearly, a complete chemical structure cannot be represented using basic SQL data types, such as numeric or text. While a compound name (text) might be considered a good representation of molecular structure, there are better ways to represent molecular structure. In this chapter, a compound will simply be identified with a unique compound id serving as a foreign key to a more complete representation of structure in another table. The following chapters will show ways that chemical structures can be fully integrated into the tables of a database. This chapter concentrates on the proper use of chemical data and ways in which multiple relational tables can be used.

The vast majority of chemical information consists of text or numerical data associated with a particular compound, or perhaps a mixture of compounds. Some chemical data cannot yet be associated with any particular structure, or has been measured for compounds whose structure is not yet known. It is important to consider these possibilities when designing a schema of tables to store chemical information.

6.3   Sample Schema for Tracking Chemical Samples

These are the operations that are required by users of this compound tracking system:

- Register a sample and record the chemical compound(s) it contains.
- Checkout the entire sample, or subsamples, recording the person (chemist), location (lab), and date and time.
- Return a sample to a location.

These are the questions that need to be routinely answered by users of this schema:

- Which sample(s) contain a particular compound?
- Where is and who has any or all samples of a particular compound?
- Where has a particular sample been since it was registered?
- Which compound(s) are contained in a particular sample?
- What is the molecular weight of a particular sample?

These are possible future requirements:

- Location may need to be expanded to identify particular shelves, cabinets, or drawers.
- Samples may need to be tracked as controlled substances.

These are the items that need to be stored in the schema:

- Sample
- Compound
- Chemist
- Location
- Time and date
- Molecular weight
- Checkout

Some items are simple; for example, molecular weight is a numeric value and time and date can be represented using the timestamp SQL data type. Other items are more complex, for example, compound location. For complex items, one must consider the components of that data that can be represented using the available SQL data types. For example, compound location could simply be stored as a text string. If compound location is needed only in reports, a text string is a good solution. However, in some applications, the location needs to be more flexibly defined. It might be necessary to include a stockroom, a laboratory, a cabinet, etc. Rather than store this in a single encoded text string, it is much better to define a set of data columns to hold this information so that it can be readily searched and updated. This defines a new complex data type composed of built-in data types, such as numeric, text, or date. Defining a table and its set of columns and defining the new data type does this. A general rule is to avoid encoding information in a single column. Rather, design a table (or new data type) to hold the individual components of the data.

Figure 6.1 shows an entity-relationship diagram for a compound-tracking schema. The table named registry defines the sample. The sample_id is used to relate other tables in the schema. This sample_id is a unique integer that may not be null and will serve as a primary key for this table and as a foreign key for other tables. In the figure, these attributes are encoded in the diagram using the asterisk (not null), PK (primary key), FK (foreign key), and PF (primary foreign key). When a sample is first entered, a new sample_id must be used. This will be enforced by the RDBMS because it is declared to be unique. For convenience, many RDBMS provide a serial data type, or sequence-generating functions, that provides the next integer in a series. However, the person or computer program responsible for making the initial entries in the registry table might also ensure that the sample_id is unique among the other sample_ids in the table. In addition, a date is stored in the column named registered and may not be a null value. Finally, a parent_sample_id is stored. This will be used when partial samples, rather than an entire container of the sample are checked out. This is still called a *foreign key* even though the primary key to which it is related is contained in the same table. Initially, the parent_sample_id will be set equal to the sample_id, indicating it is the primary entry for this sample. When a sample is taken, a new sample_id will be generated and its parent_sample_id will be set equal to the sample_id from which this subsample is taken. The registered date will also be stored for this subsample.

The next table to consider is the checkout table. When a sample is taken, either the parent sample or a subsample, data are recorded in the checkout table. A sample_id is recorded, the checked_out date is recorded, a valid location_id from the alocation table is chosen, and a valid chemist_id from the chemist table is chosen. Note that

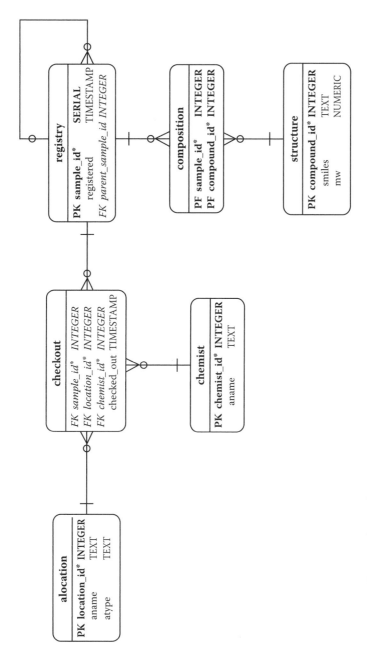

Figure 6.1 Entity relationship diagram for a chemical sample tracking schema.

the same sample _ id may be recorded multiple times in the checkout table. Each entry with the same sample _ id would have a different timestamp and likely a different location _ id and chemist _ id. This provides a record of where this sample has been from the time it was first registered to the last time it was checked out.

Why have a separate table for chemists, rather than simply storing chemist _ name in the checkout table? The alternative would be to have a column for chemist _ name in the checkout table. The advantage of storing chemist _ name in the checkout table is only that a separate chemist table is not needed. The disadvantage is that the same chemist name will be repeated in many rows of the table, possibly spelled differently or with varying upper and lowercase letters. In addition, the length of chemist _ name in each row will consist of many characters. The advantages of a separate chemist table are many. The chemist _ id is an integer, much shorter than a chemist _ name. Each chemist name is stored only once and can be corrected easily when necessary. Additional information (additional columns) about each chemist can be added to the chemist table at any time without disturbing the checkout table. Other tables can be created that relate to the chemist table if that becomes necessary. Finally, it is forbidden to store an invalid chemist _ id in the checkout table because of the foreign key constraint relating the checkout and chemist tables.

Why have a separate table for alocation related by location _ id rather than columns location_name and location _ type in the checkout table? The reasons stated above for using a separate chemist table apply here, but there are even more important advantages for the alocation table. While interviewing users of this schema, the location of each sample was seen to be of high importance. Yet, they were reluctant to define exactly what was meant by compound location and suggested that it might change in future. For example, while it might suffice today to know a sample was in lab 215 or in coldroom 12, in the future it might be necessary to know in which cabinet or which drawer a compound is located. Keeping the location information in a separate table allows the definition of location to be altered or expanded without having to modify the checkout table. A single location _ id still relates the checkout table to the alocation table.

The last two tables to consider are the astructure table and the composition table. Each sample might be composed of multiple chemical compounds, or might even be of unknown chemical composition. The astructure table has a unique entry for each compound of interest, with a unique compound _ id, a SMILES string identifying the structure of the compound, and a molecular weight. SMILES is a text representation of chemical structure that is explained more fully in Chapter 7 of this book. Other attributes of a chemical compound could also be stored in the astructure table. It is also possible to create new tables having compound _ id as a foreign key relating to the astructure table.

The purpose of the `composition` table is to provide the ability to store the information that a sample consists of one or many compounds. There might be many rows in composition, each with the same `sample _ id` and different `compound _ ids`. Or there may be a single entry in the `composition` table indicating that a sample contains only one compound. Finally, there may be no entry in composition for a `sample _ id` indicating that the sample is of unknown composition.

This schema can be expanded in many ways. For example, other information about the sample can be added, such as whether the sample is a liquid, crystal, solution, etc. If necessary, a table might be used to store the `sample _ ids` of toxic or radioactive compounds, or of compounds monitored by some governmental regulatory agency. Rather than trying to foresee all possibilities and add columns to the `sample` table, it is much simpler and more robust to add new tables as new information becomes available or necessary.

A general rule is this: Keep each table as simple as possible, with the fewest number of columns, each of which is essential to describe the entity (e.g., sample, compound, or chemist). Assign a unique integer id column and use that id in relationship to other tables containing more information or related information.

6.4 Schemas for PubChem Data

In the previous section, a schema was described for a compound tracking system based on user specifications. The designer is free to create new schemas and tables to fit the user specifications. Sometimes, an existing system needs to be analyzed in order to fit into an RDBMS model. Often, the system will have been implemented using several sets of files, with various programs implementing relationships among these files and the data in them. In this case, the structure of the schemas and tables is "suggested," or even required by the structure of the existing data.

The U.S. National Institutes of Health PubChem project contains information on millions of chemical compounds.[1] The data are divided into three main sections. PubChem Substance contains structures supplied by depositors. PubChem Compound contains unique structures with computed properties. PubChem BioAssay contains bioactivity assay results supplied by depositors. The data in these three sections are recorded independently, yet there are chemical relationships among these sections. For example, information available as a PubChem BioAssay is associated with a particular substance for which the data were collected. A substance may be a single compound or a mixture of several compounds.

In order to find structures and data in PubChem, there are search tools available online.[2] This may suffice for your needs. The data are also available in the form of SDF files and csv (comma-separated values) files.

nci_h23	
sid	INTEGER
ext_datasource_regid	INTEGER
cid	INTEGER
activity_outcome	INTEGER
activity_score	INTEGER
activity_url	TEXT
assaydata_comment	TEXT
assaydata_revoke	TEXT
log_gi50_M	NUMERIC
log_gi50_ugml	NUMERIC
log_gi50_v	NUMERIC
indngi50	INTEGER
stddevgi50	NUMERIC
logtgi_m	NUMERIC
logtgi_ugml	NUMERIC
indntgi	INTEGER
stddevtgi	NUMERIC

Figure 6.2 Entity-relationship diagram for nci_h23 data table.

This section will show how these files can be used to populate a schema of tables designed for PubChem data. While your chemical information may not correspond exactly to this schema, it should be instructive to see how the PubChem schema is designed and used.

6.4.1 BioAssay Data

PubChem BioAssay is available as hundreds of different files.[3] The files are named, for example, 1.csv.gz, 1.descr.xml, 2.csv.gz, 2.descr.xml. The xml files are descriptions of the data contained in the corresponding csv file, which results when the csv.gz file is unzipped. For example, the file 1.descr.xml contains the information: "Growth inhibition of the NCI_H23 human Non-Small Cell Lung tumor cell line is measured as a screen for anti-cancer activity" as well as information about the various columns of data in the 1.csv file. This information is used to define a table to hold the data in the 1.csv file. Figure 6.2 shows a representation of the table, named nci_h23. Using additional information in the 1.descr.xml file and using the capabilities of the RDBMS to incorporate comments on tables and columns, the following SQL defines the nci _ h23 table.

```
Create table pubchem.nci_h23
(
    "sid" Integer,
    "ext_datasource_regid" Integer,
    "cid" Integer,
    "activity_outcome" Integer,
    "activity_score" Integer,
    "activity_url" Text,
```

```
    "assaydata_comment" Text,
    "assaydata_revoke" Text,
    "log_gi50_M" Numeric,
    "log_gi50_ugml" Numeric,
    "log_gi50_v" Numeric,
    "indngi50" Integer,
    "stddevgi50" Numeric,
    "logtgi_m" Numeric,
    "logtgi_ugml" Numeric,
    "logtgi_v" Numeric,
    "indntgi" Integer,
    "stddevtgi" Numeric
);
Comment on table pubchem.nci_h23 is 'Growth inhibition of the
    NCI_H23 human Non-Small Cell Lung tumor cell line is measured as
    a screen for anti-cancer activity. Cells are grown in 96 well
    plates and exposed to the test compound for 48 hours. Compounds
    are tested at 5 different concentrations and three endpoints are
    estimated from this dose response curve: GI50, concentration
    required for 50% inhibition of growth; TGI, the concentration
    required for complete inhibition of growth; and LC50, the
    concentration required for 50% reduction in cell number. These
    estimates are done by simple linear interpolation between the
    concentrations that surround the appropriate level. If a
    compound doesn't cause inhibition to the appropriate level, the
    endpoint is set to the highest concentration tested.';
Comment on column pubchem.nci_h23."sid" is
    'PubChem Substance ID SID';
Comment on column pubchem.nci_h23."ext_datasource_regid" is 'Space
    holder for PubChem external datasource RegID, empty by default';
Comment on column pubchem.nci_h23."cid" is 'Pubchem Compound ID';
```

It is important to include comments when defining tables in a schema. Comments on only the first few columns are shown in the given example. Whenever information is available from the data source, it should be incorporated as comments. When defining tables using your own data, it is advisable to use comments to include a thorough description of each table and each column of data.

Once this pubchem.nci _ h23 table is defined, the data in the csv file can be copied into the table, typically using the SQL copy command. For example:

```
Copy pubchem.nci_h23 From Stdin Delimiter ',';
```

After unzipping 1.csv.gz and removing the first line containing column names rather than data, the file can be used with the copy command above to copy data into the pubchem.nci _ h23 table.

All of the other xml and csv data files from PubChem BioAssay can be used in a similar way to define other tables in the pubchem schema. Many of the assays use the same column names and descriptions as the above pubchem.nci _ h23 schema. This table definition can be used, changing just the table name and the comment on the table.

The structure of this table and the data in it can be viewed online using the phpPgAdmin Web application.[4] While it is possible to simply browse this table, it is more useful to search for data of interest. This applies to data in any table, of course. Simple SQL statements to carry out searches can be entered using the SQL link on the upper right. For example, the following SQL statement finds rows where the substance is considered to be active.

```
Select sid, activity_outcome, "log_gi50_M", log_gi50_ugml From
  nci_h23 Where activity_outcome = 2;
```

This may be of use, but more likely some information about the actual substances and structures is needed, not just the substance id.

6.4.2   Substances

The substances in PubChem are available as a set of sdf files. The data in these files can be read by a wide variety of programs.[5] The one most directly useful here produces a file of SQL commands to create a table and copy data into it. This sdf2sql program* is available online.[6] Using the PubChem file Substance_00000001_00025000.sdf.gz, the output of sdf2sql produces the following:

```
Create Table substance (
 Title text,
 BONDANNOTATIONS text,
 CID_ASSOCIATIONS text,
 COMPOUND_ID_TYPE integer,
 EXT_DATASOURCE_NAME text,
 EXT_DATASOURCE_REGID text,
 EXT_DATASOURCE_URL text,
 EXT_SUBSTANCE_URL text,
 GENBANK_NUCLEOTIDE_ID text,
 GENBANK_PROTEIN_ID text,
 GENERIC_REGISTRY_NAME text,
 PUBMED_ID text,
 SUBSTANCE_COMMENT text,
 SUBSTANCE_ID integer,
 SUBSTANCE_SYNONYM text,
 SUBSTANCE_VERSION integer,
 TOTAL_CHARGE integer,
 XREF_EXT_ID text);
Copy substance (
```

---

* Another approach is to store the properties from the sdf file in a separate table, rather than as columns in the substance table. This is examined more fully in Chapter 11.

```
Title,
BONDANNOTATIONS,
CID_ASSOCIATIONS,
COMPOUND_ID_TYPE,
EXT_DATASOURCE_NAME,
EXT_DATASOURCE_REGID,
EXT_DATASOURCE_URL,
EXT_SUBSTANCE_URL,
GENBANK_NUCLEOTIDE_ID,
GENBANK_PROTEIN_ID,
GENERIC_REGISTRY_NAME,
PUBMED_ID,
SUBSTANCE_COMMENT,
SUBSTANCE_ID,
SUBSTANCE_SYNONYM,
SUBSTANCE_VERSION,
TOTAL_CHARGE,
XREF_EXT_ID)
From stdin delimiter ',';
1,\N,449635 1,0,MOLI,MOLI000002,\N,\N,\N,\N,\N,\N, MOLI - NCI
    Molecular Imaging Agents\nFGCV,1,MOLI000002,1,0,MOLI000002
2,2 11 5\n20 34 5\n25 31 6\n28 55 5\n5 13 6\n58 59 6\n8 33 5,\N,0,
    MOLI,MOLI000003,\N,\N,\N,\N,\N,\N,MOLI - NCI Molecular Imaging
    Agents\n[99mTc]-P2S2-BBN(7-14),2,MOLI000003,1,0,MOLI000003
```

The column names in the Create statement are taken directly from the data tags in the input sdf file. The data types are guessed after analysis of data in the file. This sample includes only two lines of actual data from the 13,036 entries in the file.

Once this substance table is in place, it is now possible to select in a single SQL statement any substance data along with nci _ h23 data. The previous statement can be modified as follows:

```
Select
  sid, ext_datasource_name, substance.ext_datasource_regid,
  activity_outcome, "log_gi50_M", log_gi50_ugml
From
  pubchem.nci_h23
  Join pubchem.substance On substance.substance_id = nci_h23.sid
Where activity_outcome = 2;
```

Notice the use of the Join keyword and the additional table name pubchem.substance in the From clause. This is necessary because data from this table is being selected. The additional columns selected are ext _ datasource _ name and substance.ext _ datasource _ regid in the Select clause. Any columns of interest in the substance table could be selected. Note that since there is a column named ext _ datasource _ id in both tables, it is necessary to specify that the column substance.ext_datasource_regid is desired. Finally, the clause On nci _ h23.sid = substance.substance _ id indicates that these columns are related to each other and must be used in the Join.

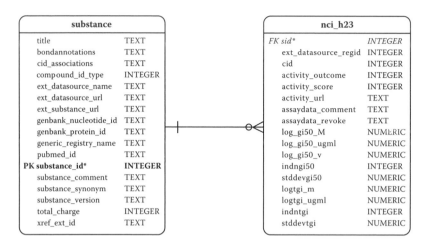

substance			nci_h23	
title	TEXT		*FK sid**	*INTEGER*
bondannotations	TEXT		ext_datasource_regid	INTEGER
cid_associations	TEXT		cid	INTEGER
compound_id_type	INTEGER		activity_outcome	INTEGER
ext_datasource_name	TEXT		activity_score	INTEGER
ext_datasource_url	TEXT		activity_url	TEXT
ext_substance_url	TEXT		assaydata_comment	TEXT
genbank_nucleotide_id	TEXT		assaydata_revoke	TEXT
genbank_protein_id	TEXT		log_gi50_M	NUMERIC
generic_registry_name	TEXT		log_gi50_ugml	NUMERIC
pubmed_id	TEXT		log_gi50_v	NUMERIC
PK substance_id*	**INTEGER**		indngi50	INTEGER
substance_comment	TEXT		stddevgi50	NUMERIC
substance_synonym	TEXT		logtgi_m	NUMERIC
substance_version	TEXT		logtgi_ugml	NUMERIC
total_charge	INTEGER		indntgi	INTEGER
xref_ext_id	TEXT		stddevtgi	NUMERIC

Figure 6.3 Entity-relationship diagram for pubchem.substance and pubhchem. nci_h23 tables.

Figure 6.3 shows the relationship between the `pubhcem.nci _ h23` and `pubchem.substance` tables in the form of an entity-relationship diagram (ERD). The primary key `substance.substance _ id` and the foreign key `nci _ h23.sid` are indicated and imply their use in an On clause when these two tables are joined.

6.4.3   Compounds

The third set of files from the PubChem repository describes chemical compounds. These are distributed as sdf files and are identified using a unique compound id. There are also multiple properties associated with each compound. Using the sdf2sql file utility described above, the table pubchem.compound is created. The compound table can then be used to locate compounds by searching any of the columns of data; for example,

```
Select * From pubchem.compound Where iupac_name Like '%aldehyde%'
And heavy_atom_count < 20;
```

would select small aldehydes. When used in conjunction with the bio-logical assay data and substance table, the compound table becomes even more useful.

From the examples in the previous section, it is clear how the sub-stance id relates `pubchem.substance` to biological assay data and how `substance` data can be selected using the substance id. How can the `compound` table be used to select compound data for substances appear-ing in one of the biological assay data tables? In other words, how is the relationship between `pubchem.substance` and `pubchem.compounds` handled?

The column `pubchem.substance.cid_associations` is taken directly from the sdf files supplied by PubChem. It has all the necessary information, but it is not in a proper form for a relation between `pubchem.substance` and `pubchem.compounds`. This is because too much information has been crammed into this column. For example, the `cid_associations` for `substance_id` 22 contains the data "449653 1 449655 2 6540406 2". This means that there are three compound ids associated with this substance id. In other words, there is a many-to-many relationship between compounds and substances. While it would be possible to parse the `cid_associations` column when the compound id is needed, it is better to have a clear relationship between substance ids and compounds ids. It is better because it enforces and preserves the relational integrity (or referential integrity) between these data. It also makes selecting data from all three data sources quicker and easier.

Another complexity is that the compound associations are classified using a small number, for example the 1 and 2 in the `cid_associations` quoted above. These classifications might be called primary when the number is 1, secondary when 2, etc. Trying to apply a parsing rule for data encoded in a column is prone to error. There is no easy way to enforce the format of data in this column.

There are several approaches to creating the relation between compound and substance. One is to create an integer column, say, `pubchem.substance.cid` that would contain only the primary compound id from the column `cid_associations`. This column becomes a foreign key related to the `pubchem.compound.cid` column. This would form a proper relation between the tables, but would neglect the secondary `cid_associations`. If those are of no interest, this approach is an excellent choice.

Another approach is to create multiple columns: one for the primary compound id and others for the secondary, tertiary, etc. compound ids. Each of these integer columns could serve as a foreign key and form a proper relation to the `pubchem.compound.cid` column. This approach is not recommended because the maximum number of compound ids in the `cid_associations` column is not known and could increase as more data is added. In addition, the type of association, primary, secondary, etc. would have to be neglected, stored in another column, or somehow encoded in the new column names. This approach has too many drawbacks to be acceptable.

The proper way to create a relation between `pubchem.substance.substance_id` and `pubchem.compound.cid` is to create a new table that acts as an intermediary. This is a typical approach to handling many-to-many relationships. This table must include a column for the compound id and a column for the substance id. There can be as many rows as necessary to provide all compound ids for any substance id. Adding a column for compound association type allows that information to be included as well. Figure 6.4 is an ERD showing how all the PubChem tables are related. The search from the previous section can now be expanded to include compound data.

```
Select
  n.sid, sc.compound_id, sc.compound_type, c.openeye_can_smiles,
  s.ext_datasource_name, s.ext_datasource_regid,
  n.activity_outcome, n."log_gi50_M", n.log_gi50_ugml
From
  compound c
  Join substance_compound sc On sc.compound_id = c.cid
  Join substance s    On s.substance_id = sc.substance_id
  Join nci_h23 n      On n.sid          = s.substance_id
Where activity_outcome = 2
Order By sid;
```

Note two additional Join clauses, each with the appropriate On clause naming the columns that relate the tables being joined. The additional columns compound _ id, compound _ type, and openeye _ can _ smiles are from the compound table. No columns are actually selected from the sub-stance _ compound table. That table is simply used to affect the many-to-many relationship between the substance and compound tables.

6.5   Data Constraints and Data Integrity

Several constraints on data have already been discussed. When a column is defined to be numeric, it is forbidden to insert anything other than a number into that column. There are constraints of this type on every standard SQL data type. This ensures a type of data integrity. When data is selected from a column with data type timestamp, the user of the data, whether a person or a computer program can be sure the data represents a valid timestamp. This type of data constraint also prevents errors from creeping into the database due to errors on data input. For example, the string 'Nan' would not be allowed in a column of type numeric. While some computer systems and languages freely use the string 'Nan' to represent (on output) "not a number," an RDBMS would reject this and never allow anything other than a valid number (or possibly a null) in a numeric column.

Another commonly used constraint is the uniqueness constraint. In previous examples, the column compound _ id was defined to be a unique integer. When the uniqueness constraint is used in a table holding a collection of compounds, it ensures that there can never be more than one compound with a particular compound _ id. This is essential if other data about a compound are stored in other tables that use compound _ id as a foreign key. Notice that this does not prevent two identical compounds

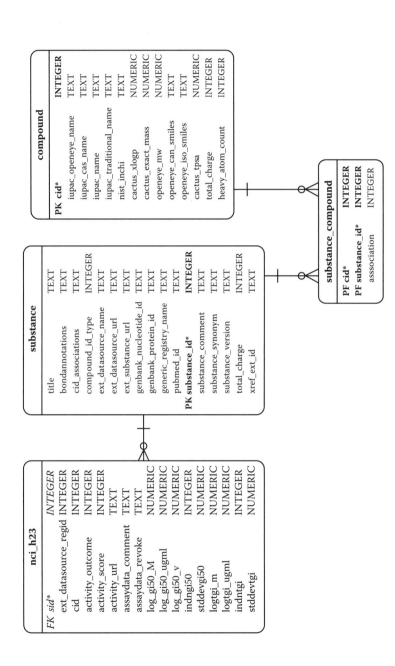

Figure 6.4 Entity-relationship diagram for all three sets of PubChem data, showing primary and foreign keys relating compounds, substances, and biological assay data.

from being entered, each with a different compound _ id. The next chapter shows how using uniqueness constraints and SMILES can ensure that a compound occurs only once in a table.

The foreign key constraint is used routinely in a schema (or schemas) with many related tables. Usually one table is defined to hold the primary unique key, for example, compound _ id. Other tables are then defined using compound _ id as a foreign key. This requires that every compound _ id in the foreign key table have a corresponding entry for that compound _ id in the primary key table. This use of primary and foreign keys forms the foundation on which the table relations and joins among relational tables are defined. The foreign key constraint can also ensure that data in the primary table cannot be accidentally deleted, since this would leave the rows in the foreign key table(s) orphaned with no corresponding primary key. The foreign key constraint does not require that the values be unique. There may be many rows with the same compound _ id in a table where compound _ id is defined as a foreign key. For example, consider a table ic50 with compound _ id as a foreign key. There may be many measures of ic50 for the same compound _ id all stored in the same table ic50. There may also be many other tables of experimental and theoretical data using the same compound _ id to store various data.

It is also possible to define other constraints on data in tables. For example, molecular weight might be stored as a numeric value. While the standard SQL numeric constraint applies, it might be desirable to also constrain molecular weight values to be positive values. This can be readily done after the column is defined using the following SQL.

```
Alter Table compound Add Constraint mw_check Check (mw > 0);
```

More elaborate constraints are also possible. For example, the chemical abstracts (CAS) number is composed of three integers. In order for a CAS number to be valid, the first integer must have no more than six digits, the second must have two digits, and the third integer is a checksum. For example, the CAS number for caffeine is 58-08-2. The values 58 and 08 are arbitrarily assigned by Chemical Abstracts, and the 2 is the proper checksum computed as $(8*1 + 0*2 + 8*3 + 5*4) \% 10$, where % represents the modulo function. The following function will validate a CAS number based on input of the three integers.

```
Create Function valid_cas(integer, integer, integer) Returns boolean
 As 'Select $1 < 1000000 And $2 < 100 And $3 < 10 And
 $3 = (
 ($2 % 10 * 1) +
 ($2 / 10 * 2) +
 ($1 % 10 / 1 * 3 ) +
 ($1 % 100 / 10 * 4 ) +
```

```
($1 % 1000 / 100 * 5 ) +
($1 % 10000 / 1000 * 6 ) +
($1 % 100000 / 10000 * 7 ) +
($1 % 1000000 / 100000 * 8 )
) % 10;' Language SQL;
```

This function, in conjunction with a CAS number-parsing function can be used to define a check constraint on a column of CAS numbers as follows.

```
Alter Table compound Add Constraint cas_check
  Check (valid_cas(caspart(cas,1), caspart(cas,2), caspart(cas,3));
```

The definition of `caspart` is not shown here, but could be written in any language supported by PostgreSQL. The `caspart` function would parse the CAS number into each of its three integer parts.

The point of this exercise is not to emphasize any particular importance of CAS numbers. In some applications, they may be stored simply as a string for reference, or even be of no interest at all. On the other hand, a corporate `compound_` id number may be very important. Often corporate ids are compound strings, for example, encoding the occurrence of various salts, or the acquisition of the compound from some external source. Functions similar to those shown above would be required to implement a check constraint.

It is a good idea to apply sensible constraints on data in order to ensure data integrity. These constraints can prevent errors and simplify the processing of data stored in the database. While there is some overhead using a check constraint, it applies only when the data are inserted or updated.

6.6 Developing Complex SQL

Many SQL statements are simple. Some statements can grow in size and become slightly more complex. For example, one might select dozens of columns from a table using a sizable, but simple `select` statement. Some of the columns might include a `where` clause, but this is not the kind of complexity that requires careful thought when constructing the SQL statement. On the other hand, a `select` statement can become quite complex, involving joining many tables with associated `where` clauses. It is very common to forget a `join` condition resulting in many more rows than anticipated. Rather than try to write complex SQL statements in one attempt, it is worthwhile to approach this systematically, as if writing a computer program.

Traditional computer languages combine sequential shorter statements or program lines to produce a result. Of course, it is possible to write such functions using SQL or using a variety of procedural languages such as plpgsql or plperl. But even a single SQL statement can become complex enough that it requires "writing" as if it were a function or program.

Rather than individual program statement or lines, SQL combines clauses into one statement to produce a result. A simple `select` statement might look like this:

```
Select id,smiles,mw From atable Where mw < 500;
```

This would return the `id`, `smiles` and molecular weight for the compounds of interest. One added complexity might require the use of parentheses to specify the correct match. For example:

```
Select id,smiles,mw,logp From atable Where
(logp > 0 And mw < 500) Or
(logp < 0 And mw < 580);
```

This kind of complexity is straightforward and will not be considered further.

When the desired data is in two different tables, a `join` is required. It is often helpful to begin to develop complex SQL statements by considering one table at a time. For example, data from the `nci _ h23` table of pubchem schema was considered earlier in this chapter. The experimental data to be selected from that table was the `substance id` (called sid in the table), `activity _ outcome`, `log _ gi50 _ M` and `n.log _ gi50 _ ugml`. This is accomplished by the simple SQL

```
Select
 n.sid, n.activity_outcome, n."log_gi50_M", n.log_gi50_ugml
From
 nci_h23 n
Where activity_outcome = 2
Order By sid;
```

Notice that a table alias n is used to refer to the table `nci _ h23`. This shorthand notation will make it easier to express complex SQL when more and more table joins are used. This statement is spread over several lines with indentation in order to locate and modify SQL clauses more easily.

Along with these experimental results, information about the substance is desired. This information is in the substance table, indexed by `substance _ id`. The SQL statement discussed previously is modified as follows.

```
Select
 n.sid, n.activity_outcome, n."log_gi50_M", n.log_gi50_ugml,
 s.ext_datasource_name, s.ext_datasource_regid
From
 nci_h23 n
 Join substance s On s.substance_id = n.sid
Where activity_outcome = 2
Order By sid;
```

Finally, data about the compounds for each substance is also needed. The compound _ id is contained in the substance _ compound table. The SQL is extended to become

```
Select
 n.sid, n.activity_outcome, n."log_gi50_M", n.log_gi50_ugml,
 s.ext_datasource_name, s.ext_datasource_regid,
 sc.compound_id, sc.compound_type
From
 nci_h23 n
 Join substance s On s.substance_id = n.sid
 Join substance_compound sc On sc.substance_id = s.substance_id
Where activity_outcome = 2
Order By sid;
```

The data about the compounds is contained in the compounds table that is indexed by compound _ id. The SQL now becomes

```
Select
 n.sid, n.activity_outcome, n."log_gi50_M", n.log_gi50_ugml,
 s.ext_datasource_name, s.ext_datasource_regid,
 sc.compound_id, sc.compound_type,
 c.openeye_can_smiles
From
 nci_h23 n
 Join substance s On s.substance_id = n.sid
 Join substance_compound sc On sc.substance_id = s.substance_id
 Join compound c On c.cid = sc.compound_id
Where activity_outcome = 2
Order By sid;
```

Notice that each line in the select clause contains columns from only one table. Likewise, each line in the from clause contains one new table name with each table (after the first one) preceded by the join keyword. Each column uses a table name (a table alias) for brevity. Now that all the tables are properly joined, the columns selected can be arranged in any order desired. Some columns can also be removed from the select clause. The compound _ id and sample _ id may not be of interest in the final result, since these arbitrary values are used only to maintain relations among the tables. Finally, any additions to the where clause can be added as desired.

The above approach works well when tables are joined using keys intended to relate tables to each other, as compound _ id and substance _ id here. The use of the join ... on construct is typical of tables joined using keys. Sometimes tables are intentionally joined without an on condition. When this is done, every row of each table is combined with every row of the other, potentially producing a large number of rows being selected. Unless you intend to join every row of each table, be sure each `join` clause contains an on condition.

There are other ways to construct an SQL statement that will select exactly the same rows and columns from these tables. These are just syntactical differences or stylistic differences. The methods shown here is just one suggestion. Depending on which RDBMS is used, different approaches may be more or less efficient.

6.7   Subselect Statements

In the SQL examples discussed previously, tables were joined with each other using the on condition to correlate the appropriate rows and a final where clause to restrict the selection of data. Without using the on condition, every row of one table would be joined with every row of the other, resulting in more rows than desired. Sometimes, one wishes to join all rows from one table with all rows from another to result in all possible combinations of rows. Unless the tables are relatively small, this may still result in more rows than desired. For example, in a table of `nci.structures` containing only 250,000 structures, combining all rows with each other would result in 62,500,000,000 rows! Even if a where clause is used to restrict the number of selected rows, it is inefficient (and unnecessary) to produce combinations in this way.

For example, one may wish to combine amines and carboxylic acids for consideration in a combinatorial chemistry experiment. The following SQL would produce 96 rows.

```
Select amine.smiles As amines, acid.smiles As acids
 From nci.structure amine, nci.structure acid
Where matches(amine.smiles, 'C[N!H0!R][C;D4]')
And matches(acid.smiles, 'CC(=O)[OH]')
Limit 96;
```

However, each of the 96 rows contains the same amine. Table 6.1 is a subset of the rows resulting from the above SQL.

What might be desired instead is a test set of 8 amines and 12 acids for a total of 96 rows. This can be accomplished if the amines are selected separately from the acids, each in a select statement of their own. The following SQL will accomplish this.

```
Select amine.smiles As amines, acid.smiles As acids From
(Select smiles From nci.structure
 Where matches(smiles, 'C[N!H0!R][C;D4]') Limit 8) amine,
(Select smiles From nci.structure
 Where matches(smiles, 'CC(=O)[OH]') Limit 12) acid;
```

Table 6.1 Selected Amines and Acids from nci.structure Tables

Amines	Acids
CC(C)(C(=O)O)NC(=O)N(CCCl)N=O	c1cc(oc1)C(=O)C(=O)O
CC(C)(C(=O)O)NC(=O)N(CCCl)N=O	CC(CCC(=O)O)N
CC(C)(C(=O)O)NC(=O)N(CCCl)N=O	CCC1C2CCC3=CC(=O)CCC3C2CCC1(C)C(=O)O
CC(C)(C(=O)O)NC(=O)N(CCCl)N=O	C1CC(OC1)CCC(=O)O
CC(C)(C(=O)O)NC(=O)N(CCCl)N=O	CC(C)(C(=O)O)NC(=O)N(CCCl)N=O
CC(C)(C(=O)O)NC(=O)N(CCCl)N=O	CCNC(CC(=O)N)C(=O)O
CC(C)(C(=O)O)NC(=O)N(CCCl)N=O	CCc1ccc(cc1)NC(=O)C=CC(=O)O
CC(C)(C(=O)O)NC(=O)N(CCCl)N=O	C(=O)(C(=O)O)N
CC(C)(C(=O)O)NC(=O)N(CCCl)N=O	C=COC=C.C(=CC(=O)O)C(=O)N
CC(C)(C(=O)O)NC(=O)N(CCCl)N=O	CC(=O)O
CC(C)(C(=O)O)NC(=O)N(CCCl)N=O	CCCCCCC(C(=O)O)N
CC(C)(C(=O)O)NC(=O)N(CCCl)N=O	c1cc(cc(c1)Cl)C(C(=O)O)N
CC(C)(C(=O)O)NC(=O)N(CCCl)N=O	CC(C(=O)O)NC(=O)C(F)(F)F
CC(C)(C(=O)O)NC(=O)N(CCCl)N=O	CC(C)CC(C(=O)O)N.C(CC(C(=O)O)N)CN
CC(C)(C(=O)O)NC(=O)N(CCCl)N=O	CC(CCC(C(=O)O)O)N
CC(C)(C(=O)O)NC(=O)N(CCCl)N=O	COc1ccc(cc1CC(=O)O)F
CC(C)(C(=O)O)NC(=O)N(CCCl)N=O	CCOC(=O)NNCC(=O)O
CC(C)(C(=O)O)NC(=O)N(CCCl)N=O	c1ccc(cc1)COC2CCC(CC2)(CC(=O)O)C(=O)O
CC(C)(C(=O)O)NC(=O)N(CCCl)N=O	CCCC(C(=O)NCC(=O)O)NC(=O)OCc1ccccc1
CC(C)(C(=O)O)NC(=O)N(CCCl)N=O	c1ccc2c(c1)CN3C(=O)CCC3(C2=O)CCC(=O)O
CC(C)(C(=O)O)NC(=O)N(CCCl)N=O	C(C(C(C(=O)O)N)O)O

Table 6.2 shows the first few rows selected by this SQL statement. This is the result that was desired: a total of 96 compounds consisting of a combination of 8 amines and 12 acids. These separate select statements are typically called subselect clauses of an SQL statement. They are enclosed in parentheses and named uniquely. They function as if they were tables themselves, but are actually subsets, or subselects of a table. There can be any number of subselects in an SQL statement and the subselect clause itself can be more complex than shown above. For example, one might also restrict the selected amines by molecular weight, vendor, or other criteria.

6.8   *Views*

As discussed in Chapter 3, a view is a subset of a table defined by a `select` statement. This is quite similar to the subselect statement discussed above. Such subselect statements are sometimes also called *in-line views*. Here we discuss the use of views as a persistent way to store subselect statements for use in SQL statements.

For example, a `training _ set` and a `testing _ set` could be selected at random from a single table of values as follows:

Table 6.2 Amines and Acids from nci.structure Tables Selected in Groups of 8 Amines and 12 Acids. Only three acids are shown in this truncated table.

Amines	Acids
CC(C)(C(=O)O)NC(=O)N(CCCl)N=O	c1cc(oc1)C(=O)C(=O)O
c1cc(ccc1C(=O)NC2(CC2)N3CCOCC3)Cl	c1cc(oc1)C(=O)C(=O)O
CCC(C)(NC)P(=O)(OCC)OCC	c1cc(oc1)C(=O)C(=O)O
Cc1ccccc1N=C(C(=C)Cl)NC(C)(C)C	c1cc(oc1)C(=O)C(=O)O
CNC1(CCCCC1)C(=O)O	c1cc(oc1)C(=O)C(=O)O
CCOC(=O)NC(C)(C)CO	c1cc(oc1)C(=O)C(=O)O
CC(C)(C)NCCS	c1cc(oc1)C(=O)C(=O)O
CC(C)C(C(=O)OC)(NC(=O)C)S	c1cc(oc1)C(=O)C(=O)O
CC(C)(C(=O)O)NC(=O)N(CCCl)N=O	CC(CCC(=O)O)N
c1cc(ccc1C(=O)NC2(CC2)N3CCOCC3)Cl	CC(CCC(=O)O)N
CCC(C)(NC)P(=O)(OCC)OCC	CC(CCC(=O)O)N
Cc1ccccc1N=C(C(=C)Cl)NC(C)(C)C	CC(CCC(=O)O)N
CNC1(CCCCC1)C(=O)O	CC(CCC(=O)O)N
CCOC(=O)NC(C)(C)CO	CC(CCC(=O)O)N
CC(C)(C)NCCS	CC(CCC(=O)O)N
CC(C)C(C(=O)OC)(NC(=O)C)S	CC(CCC(=O)O)N
CC(C)(C(=O)O)NC(=O)N(CCCl)N=O	CCC1C2CCC3=CC(=O)CCC3C2CCC1(C)C(=O)O
c1cc(ccc1C(=O)NC2(CC2)N3CCOCC3)Cl	CCC1C2CCC3=CC(=O)CCC3C2CCC1(C)C(=O)O
CCC(C)(NC)P(=O)(OCC)OCC	CCC1C2CCC3=CC(=O)CCC3C2CCC1(C)C(=O)O

```
Select logp From properties Where md5(logp) > md5(logp+1);
Select logp From properties Where md5(logp+1) < md5(logp);
```

Here, the md5 function is a hash function available in PostgreSQL. It is used as a method to partition the logp values in the properties table into two arbitrary sets of about the same size. The less than operator ensures exactly two sets, and the use of the md5 function ensures that the sets are arbitrary and of about the same size. Note that using md5 results in arbitrary but not random sets. In other words, each time the select statements above are run, exactly the same sets will result, as long as no new rows are inserted. Rather than use this SQL statement every time the test set is desired, a test _ set view and training _ set view can be defined as:

```
Create View test_set As Select smiles, logp From properties
 Where md5(logp) > md5(logp+1);
Create View training_set As Select smiles, logp From properties
 Where md5(logp+1) < md5(logp);
```

The view test _ set and training _ set can now be used as if they were actual tables. If there are other criteria desired to define a test set or training set, those can be used in the definition of the view, in place of or in addition to the md5 function used above.

A view is not limited to a subset of any one table. A view can be created using a complex SQL statement involving a join of multiple tables. Using the pubchem example above, a view might be created as:

```
Create View nci_h23_set1 As Select
 n.sid, n.activity_outcome, n."log_gi50_M", n.log_gi50_ugml,
 s.ext_datasource_name, s.ext_datasource_regid,
 sc.compound_id, sc.compound_type,
 c.openeye_can_smiles
From
 nci_h23 n
 Join substance s On s.substance_id = n.sid
 Join substance_compound sc On sc.substance_id = s.substance_id
 Join compound c On c.cid = sc.compound_id
```

This view can be used to simplify selections. For example, the SQL statement

```
Select compound_id, "log_gi50_M" From nci_h23_set1
 Where activity_outcome = 2 Order By sid;
```

is much easier to read (and maintain) than the analogous, lengthy statement in the previous section of this chapter. Even if the entire set represented by the view nci _ h23 _ set1 is never used, the definition of the set is very useful. The definition of the view can be changed at any time to allow for more columns from the original tables or to accommodate any other change to the definition of the view.

The view itself is not stored as a copy of the subset of the table. Rather, the view is a dynamic representation of the subset that changes as rows of the corresponding tables are updated, inserted, or deleted. A view is analogous to a program or a function that is executed when necessary to provide a result. It is stored in a schema in the database and can be used anywhere in an SQL statement that a table can be used. However, it is not possible to insert rows into or to update rows of a view. Instead, the original table or tables containing the data must be updated.

When a view is used, the selection contained in the view is executed each time the view is used. This can be time-consuming. It is possible to create a new table that is a real copy of the view. This is sometimes called a *materialized view*. This can speed up SQL statements that use the view. There is an automated procedure in Oracle to maintain materialized views, with the frequency of the copy set by the user. There is currently (April 2008) no automatic procedure in PostgreSQL to maintain materialized views, although there are suggestions for how to do this.[7]

References

1. PubChem Compound. http://www.ncbi.nlm.nih.gov/sites/entrez?db= pccompound (accessed April 18, 2008).
2. PubChem Text Search. http://pubchem.ncbi.nlm.nih.gov/ (accessed April 18, 2008).
3. PubChem BioAssay. http://www.ncbi.nlm.nih.gov/sites/entrez?db= pcassay (accessed April 18, 2008).
4. What is phpPgAdmin. 2008. http://phppgadmin.sourceforge.net/ (accessed April 18, 2008).
5. O'Boyle, N., 2007. Pybel—Hack that SD file. http://baoilleach.blogspot. com/2007/07/pybel-hack-that-sd-file.html (accessed April 18, 2008).
6. O'Donnell, T.J., 2008. Sdf2sql program. http://www.gnova.com/software. html (accessed April 18, 2008).
7. Materialized views. http://snapshot.projects.postgresql.org/ (accessed April 18, 2008).

## chapter 7

Computer Representations of Molecular Structures

7.1 Introduction

There are many ways to represent molecular structures. A drawing is perhaps the most common and useful. It is easy to store drawings of molecules in a computer, but a stored drawing does not constitute a useful computer representation. What is needed is a computer representation that allows structures to be stored and searched in chemically useful and meaningful ways. This is sometimes accomplished by storing a connection table containing atom and bond information. Additionally, two- or three-dimensional coordinates are often stored. These data can be stored in files or data structures in some computer language. This chapter introduces ways of representing molecular structures that take advantage of the relational model of data in a relational database management system (RDBMS).

According to standard valence bond theory,[1] the essential components of a molecular structure are the atoms and bonds composing the molecule. Two atomic properties are required—the atomic symbol or atomic number and the formal charge on the atom. Other atomic properties might be useful in many applications but are not considered as essential for the discussion in this chapter. The two atoms participating in a bond define the bond. A bond type, namely single, double, or triple, completes the definition of a bond. It is frequently convenient to use the notion of aromaticity[2] to further classify bonds and even the atoms composing that bond. The concept of a hydrogen bond is not considered here.

Files, such as SDF molfiles[3] or PDB[4] files are commonly used to represent molecular structures. The data in these files contain information about the atoms, atom charges and aromaticity, and bonds between the atoms. It is possible to define a relational table where each of the data fields in the file is stored in a separate column. One could write structured query language (SQL) to store and search data in such tables, but there is a more succinct way to represent the same information.

7.2 SMILES Representation of Molecular Structure

SMILES (Simplified Molecular Input Line Entry System) was invented by Weininger[5] to facilitate the representation and manipulation of molecular structures using computers. It uses standard atomic symbols to represent atoms and the symbols – for single bond, = for double bond, and # for triple bond. Hydrogen atoms can be represented explicitly but are almost always represented implicitly using normal conventions of valence bond theory. Single bonds need not be explicitly written. For example, propane is C–C–C or simply CCC. Methylamine is CN, and C#N is hydrogen cyanide. Propene is C=CC. For more complex structures with branched bonds, parentheses are used. For example, CC(C)O is isopropyl alcohol, whereas CCCO is propanol.

Notice that there are several ways in which SMILES could be written for the same structure, even the simplest ones. For example, hydrogen cyanide can be written as C#N or N#C, propene is either C=CC or CC=C. More complex structures can have three or many more SMILES that represent the same structure. If there were one standard way to write SMILES, then standard SQL text comparisons could be used to locate any particular structure. SMILES would become a uniquely spelled "name" for each unique structure. Canonical SMILES does just that. Using rules about which atoms should come before other atoms in the spelling of each SMILES, a unique name for each molecular structure can be provided.[6]

Once there is a unique, canonical SMILES available, this can be stored in a text column and a direct lookup for a specific structure can be done using the SQL = operator. If canonical SMILES is stored in a text column named `cansmi`, one can locate isopropyl alcohol using the SQL clause `Where cansmi = 'CC(C)O'`. And because text data can be indexed in SQL, this lookup is extremely fast. In addition, SQL uniqueness constraints can be used to enforce data integrity when using canonical SMILES.

The rules for canonical SMILES are complex and not further discussed here. There are many computer programs and structure-drawing applications that recognize and produce SMILES and canonical SMILES. There are also many programs that can interconvert molecular structure files and SMILES. To make full use of canonical SMILES in relational tables, it is not sufficient to use external programs such as these to process SMILES. There needs to be a way to integrate SMILES processing into the database and into SQL itself. This can be accomplished using SQL extensions.

7.3 Extensions to SQL for Chemical Structures

Standard SQL data types, such as integers, float, and text, are useful for storing scientific data, such as counts, measurements, and names.

Operators, such as +, *, || and functions such as sqrt, round, and upper can be used with these data types. SQL has the ability to search data, using functions such as =, <, and the like. The goal of the SQL extensions is to enable SMILES to be handled as readily as any standard data type. This requires that SQL be extended to validate and standardize, or canonicalize SMARTS. In addition, these SQL extensions provide functions and operators to allow comparisons and searches of molecular structures stored as SMILES.

Complete molecular structures can be stored as canonical SMILES in a text column. A structure can be located using an SQL clause such as where cansmi = 'CC(C)O'. But, writing an SQL clause like this requires knowing the standard way of spelling the canonical SMILES for the search structure. External programs could be used to generate the canonical SMILES, but extending SQL to generate the canonical SMILES is a better approach. First, consider how text data is typically used in a relational database.

Suppose a text column is used to store researcher's names. A good database design requires that data be represented in some standard way. In this case, upper and lower case standardization might be used. One common standard requires that the data be stored in lowercase. When an SQL search clause is written, the lowercase string would be used. For example, where name = 'einstein' would find rows storing data about einstein, but where name = 'EINSTEIN' would find no rows. It is burdensome to rely on users or programmers to always type the search name in lower case. When taking input from various sources, mixed upper and lower case data will invariably be encountered. The lower function of SQL makes using the lowercase standard easy. For example, Where name = lower('Einstein') or Where name = lower('EINSTEIN') would each find the intended rows.

If canonical SMILES are stored in a text column, a direct lookup is possible. But relying on users or programmers to know the canonical spelling of every potential search query is not advised. To use canonical SMILES in a way analogous to the lowercase example above, a cansmiles SQL function is proposed. Using this function, it becomes possible to use either the clause where cansmi = cansmiles('C#N') or where cansmi = cansmiles('N#C'). Either of these SQL clauses would find the hydrogen cyanide rows in a table that stores canonical SMILES in a column named cansmi. A new function, such as the cansmiles function described here is commonly called an SQL extension, since it extends the capabilities of standard SQL. The cansmiles (or equivalent) function is available in database extensions from gNova,[7] Daylight,[8] and others.[9] The Appendix of this book shows how to create a cansmiles extension function using the open source database PostgreSQL and open source python or perl modules. A related function, named valid returns true or false depending on whether the smiles argument is a valid smiles regardless of whether it is canonical or not.

The `cansmiles` function should also be used to insert each SMILES when a table is created. For example:

```
Insert Into structures (cansmi) Values cansmiles('CC(O)C');
```

This ensures that the same standardization is used for storing the data and for searching the data. It is not sufficient to rely on the various external programs that can read and write canonical SMILES. Each program will have a canonical SMILES method that is self-consistent, but it is likely not identical to other programs' methods. There is, unfortunately, no universally agreed-upon method to produce canonical SMILES. Once one method is chosen to implement the `cansmiles` SQL extension, it is essential for data integrity to use that method for all database operations requiring canonical SMILES.

The `cansmiles` function can also be used to enforce an SQL constraint that the `cansmi` column must contain valid canonical SMILES. SQL constrains like this are commonly used to maintain data integrity. For example, the SQL clause `check (cansmi = cansmiles(cansmi))` can be used in the initial creation of the table. One might also consider using an SQL trigger to handle an insert or update to a column that is required to contain canonical SMILES.

If any structures contain stereochemical atomic centers, consider using the `isosmiles` function instead of the `cansmiles` function. The `isosmiles` function and isomeric SMILES are discussed in a later section of this chapter.

Of course, it is possible to use any SMILES to represent a structure instead of the canonical SMILES. This makes it easier to use various external methods and programs for creating or drawing input SMILES. But unless canonical SMILES is used, the direct lookup capability is lost, or at least made less efficient. For example, one could store any SMILES in a text column named `smiles`. A search using the SQL clause `where cansmiles(smiles) = cansmiles('CC(O)C')` would work just fine, but is less efficient than storing the canonical SMILES in the first place. In some cases, it is desirable to store other SMILES spellings of each structure in addition to the canonical SMILES. This is a perfectly good practice, but these additional SMILES columns should be considered as alternate spellings of the standard canonical SMILES.

7.4 SMARTS Representation of Molecular Searches

Using canonical SMILES is a very powerful technique for molecular structure storage and lookup. However, it is sometimes necessary to perform a substructure search and not just a direct structure lookup. For example, the simplest molecule containing the cyano group is hydrogen cyanide. Although the canonical SMILES for hydrogen cyanide is C#N and for acetonitrile is CC#N, it is not possible to find all structures that contain a cyano group simply using the SQL clause where smiles like '%C#N%'. In any one SMILES, the cyano group is not always spelled C#N. For example, the canonical SMILES for thiocyanic acid is C(#N)S and not SC#N, even though both C(#N)S and SC#N represent thiocyanic acid.

Another SQL extension is needed that can understand the molecular structural nature of the SMILES string and treat it like more than just a text string. Suppose there is a function matches(A,B) that returns true when structure A contains structure B. Both these structures could be represented as SMILES and the matches function itself would understand the molecular nature properly. Then matches('C(#N)S', 'C#N') would be true as would matches('SC#N', 'N#C'), as intended. The matches function can be used to find all cyano-containing structures in a table using an SQL clause such as where matches(cansmi, 'C#N').

Sometimes the desired substructure is not as simple as a cyano group. For example, to search for di-halogen-substituted carbons, one could use an SQL clause where matches(cansmi, 'FCF') or matches(cansmi, 'FCBr') or .... This would continue this for all possible combinations of all the halogens. This is tedious. Weininger[10] proposed yet another language, SMiles ARbitrary Target Specification (SMARTS), to succinctly specify substructural searches. A SMARTS for di-halogen-substituted carbon is [F,Cl,Br,I]C[F,Cl,Br,I]. The comma-separated atomic symbols within brackets allows any one of the atoms in the list. So the SQL clause where matches(cansmi, '[F,Cl,Br,I]C[F,Cl,Br,I]') will accomplish the search for di-halogen substituted carbons. There are many other operators and symbols defined for SMARTS. These allow specification of the hydrogen atom count, heavy atom count, charge, bond types, and other aspects of atoms and bonds in substructure searches.

The matches(A, B) function is properly defined having A represent a structure using SMILES and B represent substructures using SMARTS. Of course, B may also be a SMILES.* In this case, matches will be true when B is a substructure of A. All structures in a table for which CC(O)C is a substructure can be found by using the SQL clause where matches(cansmi, 'CC(O)C'). All these structures are properly called *superstructures*, yet the search itself is commonly called a *substructure search*, because it is a search *by substructure*. Notice what happens if the arguments are reversed as in matches('CC(O)C', cansmi). All rows having cansmi as a substructure of CC(O)C will be found. These are called fragments of CC(O)C, although they could properly be called substructures of CC(O)C.

---

* Convince yourself that any valid SMILES is also a valid SMARTS, but not vice versa.

The matches(A,B) function returns true when SMARTS B matches SMILES A. It is sometimes useful to know how many times B matches A. For this, a new function is defined: count _ matches(A,B). It returns an integer, possibly 0. For example, count _ matches('CC(O)C', 'C') returns 3. The SQL clause where count_matches(cansmi, '[F,Cl,Br,I]') > 2 will find all structures having more than 2 halogen atoms. In later chapters, examples will show how this function can be used to compute molecular properties and screen structures that conform to Lipinski's Rule of 5.[11]

Another useful SQL extension function is list _ matches(A,B). This returns an array of integers telling which atoms in SMILES A were matched by SMARTS B. For example, list _ matches('CC(O)C', 'C') returns the array {1,2,4}. This list can be used for additional processing of the matches SMILES, for example, to color the matched atoms in a drawing or viewing application.

7.5   SMILES and SMARTS Quirks

SMILES may be a friend, but like all friends they have quirks that one comes to accept. The following quirks should be carefully considered before creating a large database of structures. A simple decision made early in the design can prevent troublesome changes that might otherwise be required.

7.5.1   Hydrogen Atoms

One important issue is how SMILES and SMARTS process hydrogen atoms. SMILES is almost always used without explicitly showing the hydrogen atoms. This is possible in almost all organic structures because of the predictable valence and bonding patterns of almost all organic structures. For example, propane is CCC. It is possible to write it as C([H])([H])([H]) C([H])([H]) C([H])([H])([H]), or even [CH3][CH2][CH3], but this is almost never done because it is lengthy, requires more computer processing, and does not provide any more real information than just CCC. The situation for SMARTS is not that simple.

When CC is used as a SMILES, it means exactly ethane, exactly [CH3] [CH3]. When CC is used as a SMARTS, it will of course match ethane, but will also match any structure having a C–C single bond, regardless of how many H atoms are also bonded to each C. This may be exactly what was intended, but SMARTS can be more exact in what is meant. For example, the SMARTS [CH][CH0] will only match structures having a C–C single bond where one C has exactly one H atom and the other C has none. When brackets are used for a C atom in SMILES, the assumptions normally made about the valence and hydrogen count of the atom are not used. The SMILES [CH][CH0] is a strange molecule indeed and is likely an error if it is encountered.

A common problem arises when one uses explicit H atoms in a SMARTS. For example, the SMARTS C([H])[CH0] contains an explicit H atom. It will only match SMILES that also contain an explicit H atom. Most every database has zero such SMILES. For this reason, it is important to emphasize that the SMARTS C([H])[CH0] does not represent the same substructure as [CH][CH0]. Unless one carefully designs a database to include explicit H atoms in every SMILES, explicit H atoms should *not* be used in SMARTS. This includes uses of H in, for example, C[H,F,Cl]. This will *not* match SMILES that contain CH, unless that SMILES was stored with an explicit H atom.

7.5.2 Aromaticity

Benzene is typically thought of as a combination of two equivalent resonance structures. These could be written as the SMILES C1=C–C=C–C=C1 and C1–C=C–C=C–C=1. In order to have just one representation for benzene and other aromatic systems, SMILES handles these aromatic systems specially, treating the atoms in an aromatic ring as a special aromatic type and the bonds as a special aromatic type. The lowercase symbol is used to denote an aromatic atom in SMILES and SMARTS. The SMILES for benzene then becomes c1ccccc1. A bond between aromatic atoms is an aromatic bond, unless otherwise spelled out. For example, biphenyl can be written as c1ccccc1–c1ccccc1.

This internal aromatic handling is not done for SMARTS. For example, matches('c1ccccc1', 'C1=C–C=C–C=C1') will be false. This becomes a problem when using input from an external program, such as a sketching program that may provide SMILES or SMARTS for an aromatic system in one of the many possible resonance forms. To get around this, convert the SMILES or SMARTS using cansmiles, which will aromatize the appropriate atoms. For example, matches('c1ccccc1', cansmiles('C1=C–C=C–C=C1')) is true. However, cansmiles will fail if its input is not a proper SMILES, for example, if it contains an atom list such as [F,Cl,Br].

7.5.3 Tautomers

While the several resonance forms for aromatic systems are neatly solved using aromatic atom types in SMILES, the issue of multiple tautomers cannot be handled as neatly. This is a good thing. After all, it is quite possible to distinguish two different tautomers experimentally and measure different properties that may need to be stored in a database. On the other hand, it is not possible to distinguish two different resonance forms experimentally. Every equivalent resonance form for a structure ought to be considered to be the same structure. Resonance is simply a theoretical concept.

It is sometimes necessary to be able to recognize one structure as a tautomer of another. This could be because a user entered one tautomer and expects to find data for other tautomers, especially in cases where the tautomers are in approximately equal abundance under normal laboratory conditions. It may even be that data is stored for a compound before knowing to which tautomer the data refers. In many cases, experimental data will be measured for a mixture of tautomers, yet it will be assigned to one tautomer. There is no simple solution for handing tautomers in SMILES or in a relational database. If two or more structures are tautomers of each other, this might be recorded in another table related to the table containing the SMILES.

There are several algorithmic approaches to handling tautomers. In one approach, all possible tautomers are enumerated[12–14] based on a theoretical understanding of valence bond theory. This leads to a large number of structures, many of which are not expected to be stable or observable. This large number of tautomers of each structure would have to be stored in a database or generated when needed. Neither of these solutions seems practical. In another approach, a set of rules or transformations for commonly encountered tautomers is applied.[15,16] This leads to a smaller number of tautomers. Because they are generated from chemically known transformations, they form a more reasonable set. These two methods are useful when attempting to estimate certain physical properties of structures, such as pKa or logP.

Finally, there are algorithms available for simply recognizing when two structures are tautomers. This is sufficient to locate all isomers in a database. In general, two structures are considered to be structural isomers if they share the same molecular formula. Tautomers are a special type of structural isomer in which the connectivity of the atoms, as well as the molecular formula, is the same. For example, butane (smiles:CCCC) and isobutane (smiles:CC(C)C) are strucural isomers but not tautomers. Butyraldehyde (smiles:CCCC=O) and but-1-en-1-ol (smiles:CCC=CO) are structural isomers as well as tautomers. A direct comparison of the molecular formulae readily shows the structural isomerism. There is a text graph representation that can allow easy detection of tautomers.

SMILES is a graph representation of a molecular structure containing atom and bond information. Typically, hydrogen atoms are also suppressed and inferred by rules of typical valence states of heavy atoms. If the bond information and aromaticity of atoms are removed from SMILES, the bonding framework is preserved, but the precise electronic structure is lost. This is sometimes called a simple molecular graph to distinguish it from SMILES. For example, CCCCO is the simple graph for butyraldehyde as well as its tautomer but-1-en-1-ol. But the simple graph for butane is CCCC while that for isobutane is CC(C)C. These are not tautomers and this is shown by their different simple graphs. It should be clear that two

Table 7.1 Differences and Similarities among SMILES and
Graphs for Similar Structures

Structure	SMILES	Graph	graph.hcount
H_3C $C\,H_2$ C CH_3 $C\,H_2$	CCCC	CCCC	CCCC.H10
H_3C H CH_3 C CH_3	CC(C)C	CC(C)C	CC(C)C.H10
H_3C $C\,H_2$ $C\,H_2$ C O H	CCCC=O	CCCCO	CCCCO.H8
H_3C $C\,H_2$ H C C OH H	CCC=CO	CCCCO	CCCCO.H8
H_2C $C\,H_2$ H_2C CH_2 $C\,H_2$ CH_2	C1CCCCC1	C1CCCCC1	C1CCCCC1.H12
HC $H\,C$ HC CH C CH H	c1ccccc1	C1CCCCC1	C1CCCCC1.H6

tautomers must have the same simple graph. Yet, this is not sufficient. For example, C1CCCCC1 is the simple graph for cyclohexane as well as benzene. Yet they are not tautomers. This is because they do not contain the same number of hydrogen atoms. If the simple graph of a structure is combined with the count of the hydrogen atoms, an equal comparison of these strings will reveal when two structures are tautomers. Table 7.1 illustrates the various examples discussed here. It is useful then to have a simple graph function to help determine whether two structures are tautomers. The Appendix shows two implementations of an extension function to produce a simple graph using FROWNS/plpython and OpenBabel/plpython.

7.5.4   Valence

The valence of an atom in an organic molecular structure is almost always typical. Most all carbon atoms have valence 4, oxygen atoms 2, and nitrogen atoms 3. However, some nitrogen atoms might be represented as having valence 5. For example, nitromethane is written as CN(=O)=O showing a valence of 5 for the nitrogen atom. However, it is possible to represent nitromethane as C[N+](=O)[O-] in which nitrogen retains it "normal" valence of 3. The rules of SMILES impose no valence requirements and either representation is acceptable and correct. The rules for canonical SMILES also do not modify the valence, only the order in which the atoms appear in a SMILES.

If canonical SMILES are used in a table to facilitate direct lookup of molecular structure, it is necessary that only one unique name be used for any one structure. Similarly, if one is searching for structure-containing nitro groups, it is necessary that all nitro groups be represented using the same valence conventions. For these reason, it is essential to make a decision about the use of SMILES in certain cases, such as nitro groups. Sulfur and phosphorous atoms also must be considered carefully since they are commonly found with "unusual" valence.

It is possible to use SMIRKS transformations to modify SMILES to conform to a standard valence model. For example, if a SMILES for nitromethane is entered in the charge separated form C[N+](=O)[O-], it can be transformed to the other form CN(=O)=O. Chapter 9 discusses transformations and gives examples that will help resolve issues with structures that can be represented equally well using two distinct valence forms.

7.5.5   Chirality

It is possible to represent chirality in SMILES. This is essential to correctly define the appropriate enantiomer or stereoisomer. Many databases will contain isomers. It is possible to relate the various isomers of a structure by using their common canonical SMILES. This might be done by relaxing the uniqueness constraint on the `cansmi` column in a `structure` table, or by adding another table of stereoisomers that is related to the master table. Chirality may be used in SMARTS as well.

The `cansmiles` function will not preserve any stereochemical information in the input SMILES. This is done so that the canonical SMILES for all stereoisomers is the same. It may be preferable to keep each isomer as a unique entry in a database. The `isosmiles` function preserves the stereochemical information while also reordering the atoms in the same way as the canonical SMILES.

When searching a database, if an isomeric query is used, only structures with the identical stereochemistry will be found using either a direct lookup or the matches function. If a nonchiral query is used, the direct lookup will find matching nonchiral structures, including canonical SMILES. When a nonchiral query is used in the `matches` function, structures of all chirality will be found. There is no one best method for dealing with a database containing many chiral molecules. It is important to carefully consider how to design and search such a database.

7.5.6   Isotopes

It is possible to specify the isotope of any atom in a SMILES string. This is generally not necessary because the most common isotope is simply assumed. But if, for example, a database contains information about ^{13}C, this can be readily encoded into the SMILES using [13C] instead of simply C. The [13C] atom is considered different from the normal C atom in a SMILES. A direct lookup using canonical SMILES will not locate isotopes of the same structure. A substructure search using the `matches` function will locate isotopes. This is because the `match` function uses SMARTS to specify the desired substructure.

Isotopes can be used in SMARTS. If no isotope number is specified in SMARTS, any isotope of the atom will match. For example, `select matches ('N[13C]', 'C')` will return true. However, `select matches('SNC','[13C]')` will return false. When a specific isotope is mentioned in SMARTS, then only that isotope number will match.

7.5.7   Salts and Mixtures

Compound mixtures of structures, which include salts, may be encoded using SMILES. A period between two SMILES means that the compound SMILES represents two or more noncovalently bonded structures associated with each other, such as in a salt. For example, sodium benzoate can be represented as c1ccccc1C(=O)O.[Na], or possibly c1ccccc1C(=O)[O–]. [Na+]. It may be necessary to define a set of rules about whether to represent salts using charged atoms or neutral atoms. Even with such a rule in place, one component of this mixture may be considered the important compound and the other component the counter-ion or secondary component. In some cases, the counter-ion is obviously the smaller of the two components. This is not always true. Another approach is to define a set of typical counter-ions. This set may include large groups, such as acetate or even bigger ions. Creating a table of typical counter-ions can help identify the primary and secondary components in mixtures.

7.5.8   InChI and Canonical SMILES

Canonical SMILES is a powerful tool for encoding a molecular structure as a character string, especially for use in relational database tables. Unfortunately, there is no universally accepted algorithm for producing canonical SMILES. For example, the canonical SMILES produced by OpenEye may not be the same as that produced by Daylight, ChemAxon, or ChemDraw. This is generally not an issue, as long as the same software is consistently used for creating, storing, and searching canonical SMILES. If there were one universal canonical SMILES "name" for a molecular structure, it would be possible to use this canonical SMILES in any web document. This would greatly help lookups across the web, allowing a simple string search to find exact molecular structures.

Recently, a universal string representation method was proposed and published. The International Chemical Identifier,[17] or InChI™, is a definition and set of methods maintained by the International Union of Pure and Applied Chemistry. It promises to provide a truly universal character string representation of molecular structure. Whether it will replace the widely used SMILES is yet to be seen.

7.6   SMILES and Inorganic Structures

All the examples in this chapter have been organic structures. SMILES is not limited to storing organic structures. Every atom in the periodic can be equally well represented. However, the "organic atoms" are handled specially in SMILES. Every atom in a SMILES can be represented using the atomic symbol in brackets, for example [C], [U], or [Na]. But the atoms B, C, N, O, S, P, F, Cl, Br, and I can be used without brackets. When used without brackets, SMILES assumes the lowest normal valency for these atoms. For example, formaldehyde is written as C=O, but carbon monoxide is written as [C]=[O] or [C]=O. It could be argued that the correct SMILES for carbon monoxide is [C+]#[O−]. But this argument diverges into valence bond theory, which will not be further discussed here. See also the section above about valence in SMILES.

7.7   Other SMILES Extensions

Some external programs do not use the aromatic model for SMILES and prefer using the so-called kekule form of the SMILES. This is not a canonical SMILES but can be useful for export to a drawing program, if users prefer to see alternating double bonds in aromatic ring systems. A kekule SMILES might even be necessary for some programs, which do not handle aromatic atoms in the same way as described here. The `keksmiles` function computes one of the many valid resonance structures for an aromatic system. Of course, for structures that have no aromatic systems, the keksmiles is identical to the SMILES input to the function. This function might be used, for example to `select keksmiles(cansmi)` from a table for processing by an external drawing program.

Some external programs may also need more information about exactly how many hydrogen atoms are attached to each heavy atom. The `impsmiles` function will produce a SMILES that contains the implicit hydrogen atom count. For example, `impsmiles('CC(C)O')` returns [CH3] [CH]([CH3])[OH].

As discussed above, hydrogen atoms are handled differently from other atoms in SMILES and SMARTS. When searching for structures matching CC all structures will be found that contain ethane as a substructure. Of course, this does not mean [CH3][CH3], but rather any two single-bonded carbon atoms with any number of H atoms attached. One could be more specific and search for, say, [CH][CH] to require exactly one H atom on each carbon.

Now consider a more complex case—one where a user draws in a phenyl ring as a substructure search. The drawing program would produce c1ccccc1. If c1ccccc1 is used, any structure containing a phenyl ring will be found. The user might have intended to allow all possible substitutions in all positions on the ring, and indeed this would find those structures. If a user sketched in an R group (represented as * in SMILES), most drawing programs would produce c1ccccc1*, unless the user painstakingly set the hydrogen count on every other atom of the ring. Most likely, the user intended to require H on all positions, except the one with the *. The intended SMILES would be [cH]1[cH][cH][cH][cH]c1* instead of c1ccccc1*. To facilitate the hydrogenation of SMILES strings, the `impsmiles` function works nicely. It produces a SMILES containing all necessary hydrogen atoms, paying attention to those atoms which have a * atom attached to them. For example, `impsmiles('c1ccccc1*')` returns [cH]1[cH][cH][cH] [cH]c1*. The resulting SMILES functions very nicely as a search SMARTS for use in the `matches` function.

7.8 Input and Output of Molecular Structures

As with all data in an RDBMS, there is an external and internal representation of data. This was discussed in an earlier chapter for standard data types, such as text and numeric. For molecular structures, there is of course no SQL standard. When building a database containing molecular structures, a decision should first be made: which internal representation will be used and which external representation.

This chapter focused primarily on SMILES and canonical SMILES. It is feasible and common to use SMILES as the internal representation of molecular structure. Using the SQL functions described in this chapter, many useful features would become available, such as canonicalization, and searching. As described here, the SMILES would not truly be an SQL data type because it is actually represented as a text string. There are ways to extend SQL even further to make SMILES a data type equal in every way to other standard SQL data types. This is discussed in a later section of this chapter.

Another choice for the internal representation of molecular structure is a molfile. It would be possible to construct SQL functions like those described in this chapter that would operate on this type of data. One disadvantage of molfiles is their greater size compared with SMILES. One advantage is that it is possible to store atomic coordinates, which is not possible with SMILES. There are other molecular file formats, but these are substantially the same as a molfile, except perhaps for specific atom types that may be of use in some database applications.

The recommendation here is to use SMILES to store molecular structure itself. If other features of the molecule or atoms need to be stored, other data types and columns can be added to the row describing the molecule. It is the "SQL way" to *not* encode a lot of information into one data type. When using a molfile as the structural data type, too much data is encoded in a single data type. The individual data items must be parsed and validated. Errors creep into the data, due to missing, extra, or invalid portions of the molfile. Ways of storing atomic coordinates, atom types, and molecular properties are discussed Chapter 11.

The external representation of molecular structure is a less rigorous definition. For example, there are many programs available that can convert to and from SMILES and molfiles. These can be used when a molfile (the external representation) needs to be imported as a SMILES (the internal representation) into the database. Similarly, a SMILES can be easily exported as a SMILES or converted to a molfile or other file format. It is useful to have these conversion functions as SQL extensions.

Consider the extended SQL functions smiles _ to _ molfile and molfile _ to _ smiles. Having these functions available as SQL extensions allows one to export a molfile from a table containing SMILES. For example:

```
Select smiles, smiles_to_molfile(smiles) from atable;
```

outputs a SMILES string and a molfile as a text string. Similarly, the function molfile _ to _ smiles could be used to convert a text string representation of structure to SMILES. If the advice here is followed, a column of molfiles would not be the internal representation of molecular structure. Nevertheless, the advice here should not be construed as a recommendation against ever using molfiles. Having a column for SMILES as well as a column for molfiles will fit the needs of many database designers.

7.9 Useful SQL Extensions

Several new SQL functions have been introduced here. These functions make it possible to store molecular structures in an RDBMS as text strings. They also allow these strings to be manipulated and searched in a chemically meaningful way. This greatly expands the usefulness of an RDBMS for chemical applications. Table 7.2 summarizes these functions using SQL notation for defining functions. Much of the rest of this book will describe more useful functions and describe ways of using and extending these ever further.

The Appendix of this book shows three complete implementations of these functions using PostgreSQL and PerlMol, FROWNS, and OpenBabel modules. Each of these three modules is free and open source. Using these functions is an excellent way to become familiar with the concepts in this chapter. It is possible to extend these functions even further to take advantage of other features of PerlMol, FROWNS, and OpenBabel to satisfy the needs of many molecular modeling projects. However, each of these three modules has limitations. Before embarking on a large complex database project, a thorough examination of the limitations of PerlMol, FROWNS, and OpenBabel should be done. One important distinction between these three modules is how they generate canonical SMILES. Each one generates valid canonical SMILES, but each produces different canonical SMILES. This is simply due to differing algorithms for canonically ordering atoms. As discussed earlier, there is no universal canonical SMILES.

Table 7.2 summarizes the core functions used throughout the rest of this book. There are several commercially available chemical extensions to SQL. There may not be an exact correspondence of functions from these vendors to functions in Table 7.2.

Table 7.2 Core Chemical SQL Extension Functions

Function	Input type	Output type	Description of output
valid	Text	Boolean	Tests whether SMILES is valid
cansmiles	Text	Text	Canonical form of SMILES
isosmiles	Text	Text	Isomeric form of SMILES
keksmiles	Text	Text	Kekule form of SMILES
matches	Text,text	Boolean	Tests whether SMILES (arg #1) matches SMARTS
count_matches	Text,text	Integer	Number of times SMARTS matches SMILES (arg #1)
list_matches	Text,text	Integer array	Atoms in SMILES (arg #1), which match SMARTS
smiles_to_molfile	Text	Text	Molfile formatted string
molfile_to_smiles	Text	Text	SMILES

7.10   SMILES as an SQL Data Type

The standard SQL data type `Text` has been used to store SMILES. This is appropriate because every SMILES is a valid text string. But not every text string is a valid SMILES. Without additional information about SMILES, the RDBMS cannot enforce any rules about which text strings ought to be in a column intended to contain SMILES.

7.10.1   Domains

The SQL domain allows one to define which values are to be allowed in a particular column of a table. A domain is created by stating the underlying built-in SQL data type used to store the domain data type. In addition, a check constraint function may be used to allow or forbid certain values. This can be used to great advantage for SMILES and canonical SMILES. Using a domain improves the ability of the RDBMS to maintain the integrity of the data contained in its tables.

The following SQL defines a domain data type smiles.

```
Create Domain smiles As Text Check (valid(Value));
```

The use of the keyword `Value` is required. `Value` refers to the value of the data element, here the SMILES. Once this domain is created, it can be used as a data type in the creation of a table. For example:

```
Create Table atable (id Integer, smi smiles, mw Numeric);
```

When a value is inserted into this table, the `valid` function will be called by the RDBMS. If the function returns true, then the value will be allowed into the column smi. Otherwise, an SQL error will be reported and the value will not be allowed.

Using a `domain` like this, the smiles data type behaves much like a standard data type. When one attempts to insert an invalid number into a numeric column, an SQL error is reported and the value is not inserted. This fundamental behavior of an RDBMS is readily extended to SMILES using a `domain`.

The check constraint used in the creation of a domain is similar to the check constraint used in the creation of a table. For example, it would be possible to simply

```
Create Table atable ( id Integer, smi Text Check(valid(smi)) );
```

This would ensure that the column smi could contain only a valid SMILES. If this is the only table in which a SMILES column is used, this approach is excellent. If a SMILES column is an important part of the database and is used in many tables and functions, creating a domain to define a valid SMILES is the better solution.

It might also be useful to define a canonical SMILES domain. This could be done as follows:

```
Create Domain cansmiles As Text Check (cansmiles(Value)=Value);
Create Table ctable (id Integer, cansmi cansmiles, formula Text);
```

This is not recommended. Instead, a trigger is a better way to handle canonical SMILES.

7.10.2   Triggers

Using a domain ensures that only appropriate data can be inserted into a column. If an attempt is made to insert invalid data, an error is reported. The user is then responsible for correcting the value, if possible and trying the insert again. The SQL trigger mechanism automates this process. The following SQL will not only ensure that the cansmi column contains canonical SMILES, it will correct problems where possible.

```
Create Table ctable (id Integer, cansmi Text Check
  (cansmi=cansmiles(cansmi)), formula Text);
Create Function canonicalize() Returns Trigger As $EOSQL$
 Declare
 cansmi Text;
 Begin
 cansmi = cansmiles(NEW.cansmi);
 If cansmi != NEW.cansmi Then
 NEW.cansmi = cansmi;
 End If;
 Return NEW;
 End;
$EOSQL$ Language plpgsql;
Create Trigger canonicalize Before Insert Or Update On ctable
  For Each Row Execute Procedure canonicalize();
```

This canonicalize function uses NEW to refer to the row being inserted or updated. NEW.cansmi refers to the value under question. The canonical SMILES is computed and compared to NEW.cansmi. If they are not the same, the NEW.smi value is replaced by the canonical SMILES value and the NEW row is returned. This NEW row is used by the RDBMS in place of the original row. The create trigger command causes this operation to be put into effect in the RDBMS.

Why use the domain to define a smiles data type, but use a trigger for canonical SMILES? First, SMILES is either valid or not. It is not feasible to write a function to correct an invalid SMILES, so a trigger to do so would not be effective. A domain is defined with a simple Boolean check constraint to allow or disallow a value. This is just what is need for a smiles domain and smiles data type. On the other hand, canonical SMILES is just a type of SMILES. It is very common that a user will attempt to insert a valid SMILES into a canonical SMILES column. This should be forbidden, but there is a function, `cansmiles` that can produce a valid canonical SMILES from a valid SMILES. A trigger is a good solution to this common situation. The `check` constraint on a cansmiles column doubly ensures that only canonical SMILES is allowed into that column.

There is another reason to avoid using a cansmiles `domain`: There is an interaction between the use of domains and triggers. A domain `check` constraint takes priority over the trigger. In other words, if one attempts to insert an invalid canonical SMILES into a column defined using the cansmiles `domain`, the insert may fail, even though there is a `trigger` on the table. This is because the `check` constraint of the `domain` forbids the insert before the `trigger` is applied by the RDBMS. The recommended solution to this is to define the `cansmi` column using the `Text` data type, use a `check` constraint on that column and a `trigger` on the table containing the `cansmi` column.

In summary, the `domain` check occurs first, then the `trigger` and finally the column constraint `check`. So, whenever a `trigger` is used to attempt to correct a value being inserted or updated, a `domain` check constraint should not be used. Instead a column `check` constraint should be used.

7.11   Summary

The use of a few new SQL functions can greatly enhance the way chemical structures are used in a relational database. These functions allow the SMILES text string to store structures and canonical SMILES to create a unique text representation of a specific chemical structure. SMARTS text strings are used to search SMILES strings in a way comparable to how regular expressions are used to search ordinary text strings. Functions that convert to and from SMILES and common chemical structure file format expand the kinds of chemical data the database can handle. Finally, the use of domains, triggers, and column check constraints can improve the integrity of the data in a database.

References

1. Pauling, L. 1939. *The nature of the chemical bond and the structure of molecules and crystals,* 3rd ed. Ithaca, NY: Cornell University Press.
2. Sainsbury, M. 1992. *Aromaticity.* New York, Oxford University Press, Inc.

3. Molfiles. 2008. http://en.wikipedia.org/wiki/MDL_Molfile (accessed April 18, 2008).

4. PDB Documentation. 2008. http://www.wwpdb.org/docs.html (accessed April 18, 2008).

5. Weininger D. 1988. SMILES, a chemical language and information system, 1. Introduction to methodology and encoding rules. *J. Chem. Inf. Comput. Sci.* 28:31–36.

6. Weininger, D., Weininger, A., and Weininger, J.L. 1989. SMILES 2. Algorithm for generation of unique SMILES notation. *J. Chem. Inf. Comput. Sci.* 29:97–101.

7. O'Donnell, T.J. 2008. CHORD. http://www.gnova.com/ (accessed April 18, 2008).

8. DayCart®: Chemical intelligence for an Oracle environment. 2007. http://daylight.com/products/daycart.html (accessed April 18, 2008).

9. O'Shea, M.D. 2008. Chemoinformatics chemistry data cartridges. http://www.strychnine.co.uk/oracledatacartridges.html (accessed April 18, 2008).

10. SMARTS tutorial. 2008. http://www.daylight.com/dayhtml/doc/theory/index.html (accessed April 18, 2008).

11. Lipinski, C.A., Lombardo, F., Dominy, B.W., and Feeney, P.J. 2001. Experimental and computational approaches to estimate solubility and permeability in drug discovery and development settings. *Adv. Drug Del. Rev.* 46:3–26.

12. Sayle, R. and Skillman, G. 2002. Hooked on protonics. http://www.eyesopen.com/about/events/presentations/acs02/tsld018.htm (accessed April 18, 2008).

13. Sayle, R. and Delany, J. 1999. Canonicalization and enumeration of tautomers. http://www.daylight.com/meetings/emug99/Delany/taut_html/index.htm (accessed April 18, 2008).

14. ChemAxon. Calculator plugins. 2008. http://www.chemaxon.com/marvin/help/calculations/calculator-plugins.html#tautomer (accessed April 18, 2008).

15. ScienceServe. 2008. Tautomer. http://www.scienceserve.com/Software/molnet/tautomer/index.htm (accessed April 18, 2008).

16. Oellien F., Cramer, J., Beyer, C., Ihlenfeldt, W., and Selzer, P.M. 2006. The impact of tautomer forms on pharmacophore-based virtual screening, *J. Chem. Inf. Model.* 46(6):2342–2354.

17. Stein, S.E., Heller, S.R., and Tchekhovski, D. 2003. An open standard for chemical structure representation: The IUPAC chemical identifier. In *Proceedings of the 2003 International Chemical Information Conference*, ed. H. Collier, 131–143. Nimes, France: Infonortics.

chapter 8

Molecular Fragments and Fingerprints

8.1 Introduction

Simplified Molecular Input Line Entry System (SMILES) is a simple, yet complete description of molecular structure that considers the atoms and bonds in a molecule. Using unique canonical SMILES, an indexed table lookup of a structure can be quickly done. For example, the SQL to lookup phenol is:

```
Select cansmi From atable Where cansmi=cansmiles('c1ccccc1O');
```

When the table contains unique canonical smiles in an indexed column `cansmi`, and the `cansmiles` function provides the proper canonical SMILES for phenol, this lookup is extremely fast.

It is often necessary to find all structures that contain a given substructure. Consider the substructure search to find all structures that contain the phenol group. Using the `matches` function described in a previous chapter, the SQL to carry out such a substructure search is:

```
Select cansmi From atable Where matches(cansmi,'c1ccccc1O');
```

This cannot make use of the index on the column `cansmi`. Every row of the table must be examined to see if the `matches` function succeeds. This is a time-consuming process compared to a direct, indexed lookup.

8.2 Fragments

One way to speed up a substructure search is to use a reduced representation of molecular structure and a corresponding alternative to the `matches` function. If this reduced representation of molecular structure is sufficiently simple and if the alternative `matches` function is sufficiently fast, they can be used as a filter to quickly decide which rows need more careful examination using the full `matches` function. Other rows for which the reduced representation does not match can be quickly passed over.

One might use molecular formula as a simpler representation of molecular structure. Ignoring H atoms, the molecular formula for phenol is C6O.

Every structure containing phenol as a substructure must have a molecular formula with 6 or more C atoms and 1 or more O atoms. Structures with fewer C or O atoms can be immediately ruled out as possible matches for phenol. Of the remaining structures, there will be some that satisfy the molecular formula comparison yet do not match phenol. The more time-consuming matches function will be used only for the final determination. Overall, the process of finding substructure matches will be faster. Exactly how much faster depends on the number of rows that can be quickly ruled out using the faster molecular formula comparison. It also depends, of course, on how fast the molecular formula comparison can be done.

One way to do a quick molecular formula comparison is to store the molecular formula not as a string representation, such as C6O, but as a column of integers. Each row in a table of molecular structures would contain SMILES, but the table would also have additional columns containing the count of each atom type. These columns could be indexed to speed up the molecular formula comparison. The SQL used to search for structures containing phenol becomes as follows:

```
Select cansmi From atable Where C_count>=6 and O_count>=1
 And matches(cansmi, 'c1ccccc1O');
```

The columns C _ count and O _ count would have been precomputed when the row for each molecular structure was added to the table.

Because every molecular structure is composed of atoms, the atom counts corresponding to molecular formula form a complete set of molecular fragments. However, the atom counts are not a very discriminating filter. Another approach is to construct a set of molecular fragments that are complex enough to discriminate various structures from one another yet simple enough to be used for fast filtering before using the full matches function.

Constructing a useful set of molecular fragments requires knowledge of the types of structures that will appear in the database. This will be discussed in a later section of this chapter. First, consider how such a set of fragments can be used to filter structures during a substructure search.

8.2.1   Fragment Keys

Suppose a representative set of N fragments has been defined. A bit string* containing N bits can be used to represent the presence or absence of each fragment in any molecular structure. This alternative representation of molecular structure is called a *fragment key*. It can be used as a filter

---

* The bit or bit varying data type in standard SQL will be used in the examples in this and following chapters. Oracle does not support this data type. PostgreSQL syntax will be used.

during a substructure search. For example, the following SQL might be used to locate structures containing the phenol group.

```
Select cansmi From atable Where (fkey&key('c1ccccc1O')=fkey)
 And matches(smiles,'c1ccccc1O');
```

In this example, the function named key returns a bit string denoting the presence or absence of each of the N fragments. The column fkey contains the bit string for each structure, precomputed using key(cansmi). The SQL clause where fkey&key('c1ccccc1O')=fkey will be true only when all bits set to 1 in fkey are also set in key('c1ccccc1O'). In other words, that clause will be true only when the structure (the row with that fkey) contains all the fragments that phenol contains. That is necessary but not fully sufficient for the cansmi of that structure to match phenol. The final matches function must be used to return the proper set of substructure matches. However, since the comparison of bit strings using the & operator is much quicker than the matches function, the bit string comparison acts as a quick filter. The more time-consuming match function is evaluated only for those structures that pass the quicker bit string test. The computation of the fkey using the key function is time-consuming for a large table of structures, but it need be done only once and stored in a row with the corresponding SMILES.

The key function used above can be written using SQL, along with a table of fragments. As a simple example, the fragments shown in Table 8.1 are used. This table could be created using the following SQL.

```
Create Table fragments (description Text, smarts Text, abit Integer);
```

The column named smarts contains the SMiles ARbitrary Target Specification (SMARTS) pattern defining the fragment. The column named

Table 8.1 Simple Fragment Keys
Defined Using SMARTS

Description	Smarts	abit
Phenyl	c1ccccc1	1
Aliphatic alcohol	C[OH]	2
Alcohol	[C,c][OH]	3
Aromatic alcohol	c[OH]	4
Aliphatic ether	COC	5
Ether	[C,c]O[C,c]	6
Aromatic ether	cOc	7
Ketone	O=[CH0](C)C	8
Carboxylic acid	O=[CH0](C)[OH]	9
Aldehyde	O=[CH1]C	10

`abit` represents which bit in the `key` will be set if the structure contains that fragment. The column named `description` is a brief description of the fragment. Consider the result of the following SQL.

```
Select abit from fragments Where matches ('c1ccccc1O', smarts);
```

The result is two rows:

```
1
3
```

because phenol is matched by only two SMARTS from the table fragments, namely those with abit=1 (phenyl) and abit=3 (alcohol). A slight modification shows how a bit string can be created.

```
Select B'1'::bit(50)>>abit-1 from fragments
  Where matches('c1ccccc1O', smarts);
```

This arbitrarily assumes the final result will be 50 bits, suitable for a table of fragments having 50 or fewer rows. The result of this SQL is:

```
10000000000000000000000000000000000000000000000000
00100000000000000000000000000000000000000000000000
```

The first row is a bit string of length 50 having bit #1 set; the second row has bit #3 set. This is getting closer to the desired single value key having both bit #1 and bit #3 set. How can these rows be combined into a single bit string with bit #1 and bit #3 set? An aggregate function similar to sum would provide the correct result. But there is no standard SQL aggregate function such as this that operates on bit strings. The Appendix to this book shows the definition of such a function, called `orsum`. Using that function, the final definition of the key function is:

```
Create Function key(text) Returns bit varying As $$
  Select orsum(B'1'::bit(50)>>abit-1) from fragments
  Where matches($1, smarts);$$ Language SQL;
```

The result of `Select key('c1ccccc1O')` is the single value:

```
10100000000000000000000000000000000000000000000000
```

This is the fragment key for phenol. The `key` function can be used to compute and store values of the fragment key in tables of molecular structures. It can also be used to compute values of fragment keys for substructures to be used as a prescreen during a full substructure search using the `matches` function.

8.2.2 MACCS Keys and Other Fragment Keys

One popular set of fragments has been published by MDL.[1] It is commonly known as the MACCS public 166 keys. The Appendix shows a table of 164 rows, analogous to the fragments table defined and used in the example in the previous section. Using this table and a slight modification to the key function defined above, a `public166keys` function can be easily defined. That function is also contained in the Appendix.

Any other set of useful fragments can be created and used as a fragment key to prescreen rows in a table during a substructure match. The `public166keys` table contains entries for every element in the periodic table, although the bulk of the table is designed to distinguish various organic compounds from one another. In a database containing a majority of other types of compounds, a different set of fragment keys is appropriate. The point here is not to provide the best set of fragment keys or even to recommend one set over another, but rather to illustrate a general method for computing fragment keys using simple SQL and a relational database table to define the fragments. The advantage of this approach is that the algorithm and code to produce the fragment key is not contained in some external program. It is an integral part of the database with the fragment table clearly exposed for verification, modification, and use in other ways.

8.3 Fingerprints

Another approach for generating bit string keys does not use a table for fragments at all. Instead, it uses an algorithm to fragment each structure and record each fragment as a bit pattern. Rather than assign each fragment to a particular bit number as is done in the fragment key tables above, some method of encoding each fragment is used. One approach is to use the SMILES string that represents the fragment and apply a hash function[2] to produce a fingerprint.

One method for producing these fragments first considers each atom as a fragment of size one, similar to the molecular formula approach described above. Then considering atoms bonded to each atom produces two-atom fragments. Multiple-atom fragments are then produced. Using this approach exhaustively and following every bond of every atom would produce every possible fragment of every possible size for each structure. This would be a large number of fragments, even for reasonably sized structures. The number of bits required to store this information would be correspondingly large. At some point, the size and complexity of the bit string representation would make the prescreening process too slow to be useful. To avoid that possibility, an upper limit on the size of each fragment is imposed.

One popular fingerprint algorithm[3] produces fragments as unbranched chains of atoms. This approach is typically called a *path-based method* because the algorithm follows continuous paths of bonded atoms. Another algorithm prefers branched fragments of each atom, creating ever-expanding neighborhoods of atoms around each central atom. This is typically called a *circular fingerprint.*[4]

Regardless of the method used to fragment the structure, the hashed fingerprint of each fragment is combined with hashed fingerprints for other fragments from the same structure to produce an overall fingerprint for the structure. This bit string is used in an equivalent way to the fragment keys above to prescreen rows of structures during a substructure match. The Appendix shows two functions to compute a fingerprint bit string.

8.4   Similarity Measures

Besides using fingerprints or fragment keys as a prescreen to speed up substructure matches, they can be used in other ways. The bit patterns for two molecular structures can be compared by considering bits they have in common due to common fragments. Bits not in common are due to fragments in one structure not appearing in the other structure. There are many ways to combine the counts of common bits, differing bits, and bit string length to produce a numerical measure of the similarity of one structure to another. One popular method is called *Tanimoto.*[5] Given a fingerprint or fragment key for structures A and B, the Tanimoto index is the ratio of the number of bits A and B have in common to the sum of the number of bits set for A plus the number of bits set for B minus the number of bits in common. An SQL definition for the Tanimoto index is as follows:

```
Create Function tanimoto(bit, bit) Returns Real As
'Select nbits_set($1 & $2)::real /
(nbits_set($1) + nbits_set($2) - nbits_set ($1 & $2))::real; '
Language SQL;
```

The & (logical AND) operator and the ~ (logical NOT) operator are used along with a nonstandard SQL function nbits _ set. This function and other related similarity functions are contained in the Appendix. The suitability of fragment keys, path-based or circular fingerprints, for any particular purpose is the subject of ongoing research.[6]

8.5   Computing Fragment-Based Properties

The methods shown above to compute fragment keys can be extended to compute fragment-based properties of molecules. The use of a relational table to define the fragments makes the computation suitable to using SQL to define the function. Rather than having the fragment parameters

Table 8.2 Atomic Weights for Some
Common Atoms and Associated SMARTS

SMARTS	Weight	Symbol
[#1]	1.01	H
[#6]	12.01	C
[#7]	14.01	N
[#8]	16.00	O
[#9]	19.00	F
[#15]	30.97	P
[#16]	32.06	S
[#17]	35.45	Cl
[*;h1]	1.01	H1
[*;h2]	2.02	H2
[*;h3]	3.03	H3
[*;h4]	4.04	H4
[*;h5]	5.05	H5
[*;h6]	6.06	H6

buried in an external computer program, exposing them in a table makes it easier to maintain, verify, and expand the parameter set.

The simplest molecular property is molecular weight. The obvious fragments to use for this are atoms. It is a simple matter to define the SMARTS fragments for any atom. Table 8.2 shows the definition for a few common atoms. The full table for the first 103 atoms is shown in the Appendix.

The following function is analogous to the fragment key function above. It uses a relational table to define fragments, a function to match SMILES and SMARTS (in this case count _ matches), and an aggregate SQL function to tally the results over all matched fragments.

```
Create Function amw(character varying)
 Returns Numeric As $EOSQL$
 Select sum(weight*count_matches($1,smarts)) From amw;
 $EOSQL$ Language SQL;
```

This function is not a very efficient method to compute molecular weight compared with a compiled C program, for example. The advantage is that the function is contained within the database and is expressed using a relational table that exposes the important parameters of the computation. The following function could be used when creating a table containing SMILES, or used when necessary, for example, to compute the molecular weight of phenol.

```
Select amw('c1ccccc1O');
Select amw(smiles) from structure;
```

Another useful fragment-based function computes the polar surface area of a molecule using the method described by Ertl, Rohde, and Selzer.[7] The SMARTS and partial surface areas for the fragments described by Ertl are shown in Table A.3 in the Appendix. That table is created as

```
Create Table tpsa (psa Numeric, smarts Text, description Text);
```

The following function computes the tpsa value using the `tpsa` table.

```
Create Function gnova.tpsa(text)
 Returns Numeric As $EOSQL$
 Select sum(psa*count_matches($1,smarts)) From tpsa;
$EOSQL$ Language SQL;
```

This function could be used to add a column of tpsa to any table containing SMILES.

There are other fragment-based methods[8,9] published that could be implemented using the approach described here. As long as the fragments can be defined using SMARTS and the molecular property is a simple sum (or product or other aggregate function), a relational table of fragment values can be used. Using the `match` or `count _ matches` functions from Chapter 7, an SQL function can be easily written to compute the property value.

References

1. Durant, J.L., Leland, B.A., Henry, D.R., and Nourse, J.G., 2002. Reoptimization of MDL keys for use in drug discovery, *J. Chem. Inf. Comput. Sci.,* 42(6): 1273–1280.
2. Hash function. 2008. http://en.wikipedia.org/wiki/Hash_function (accessed April 18, 2008).
3. Fingerprints—Screening and Similarity. 2007. http://daylight.com/dayhtml/doc/theory.finger.html (accessed April 18, 2008).
4. Bender, A., Mussa, H.Y., Glen, R.C., and Reiling, S. 2004. Molecular similarity searching using atom environments, information-based feature selection, and a naive Bayesian classifier. *J. Chem. Inf. Comput. Sci.* 44(1):170–178.
5. Tanimoto, T.T. 1957. IBM Internal Report, November 17.
6. Hert, J., Willett, P., Wilton, D.J., Acklin, P., Azzaoui, K., Jacoby, E., and Schuffenhauer, A. 2004. Comparison of fingerprint-based methods for virtual screening using multiple bioactive reference structures. *J. Chem. Inf. Comput. Sci.* 44(3):1177–1185.
7. Ertl, P., Rohde, B., and Selzer, P. 2000. Fast calculation of molecular polar surface area as a sum of fragment-based contributions and its application to the prediction of drug transport properties. *J. Med. Chem.* 43:3714–3717.
8. Wildman, S.A. and Crippen, G.M. 1999. Prediction of physicochemical parameters by atomic contributions. *J. Chem. Inf. Comput. Sci.* 39: 868–873.
9. Andrews R., Craik, D.J., and Martin, J.L. 1984. Functional group contributions to drug-receptor interactions. *J. Med. Chem.* 27(12): 1648–1657.

chapter 9

Reactions and Transformations

9.1   Introduction

The use of Simplified Molecular Input Line Entry System (SMILES) as a string representation of chemical structure makes possible much of what has been discussed in earlier chapters of this book. A chemical reaction could be represented as a collection of SMILES, some identified as reactants and some as products. It is possible to define a table to do this, or perhaps use some arrays of character data types, but a syntax extension of standard SMILES allows reaction to be expressed easily. SMIRKS is an extension of SMILES and SMiles ARbitrary Target Specification (SMARTS). It is used to represent chemical transformations. SMIRKS can also be used in a transformation function to combine SMILES reactants to produce SMILES products.

This chapter describes some of the aspects of SMIRKS and shows how it can be integrated into a relational database using new structural query language (SQL) functions. It discusses ways in which chemical transformations and reactions can be used to improve the robustness and usefulness of a chemical relational database.

9.2   Reaction SMILES

Reaction SMILES is an extension to standard SMILES used to represent a specific reaction. It uses punctuation to distinguish reactants from products. For example, the reaction SMILES CC(=O)O.CN>>CC(=O)NC.O represents the reaction of acetic acid with methylamine to form N-methylacetamide plus water. As with standard SMILES, explicit H atoms are typically not shown, although they may be. For example, the same reaction can be represented as [CH3]C(=O)[OH].[CH3][NH2]>>[CH3]C(=O)[NH][CH3].[H]O[H]. The punctuation >> is used to separate reactants from products, and the period is used to separate reactants or products from each other. There are no rules in reaction SMILES that enforce correct reaction stoichiometry or other aspects of actual chemical reactions.

Reaction SMILES can be used to store and search chemical reactions using the same functions described earlier for standard SMILES. For example, `cansmiles('CC(=O)O.NC>>CC(=O)NC.O')` returns

CC(=O)O.CN>>CC(=O)NC.O and `matches('CC(=O)O.CN>>CC(=O)` `NC.O','CC(=O)O')` returns a true value. The SQL

```
Select rxnsmiles From a table Where matches(rxnsmiles,'CC(=O)O');
```

selects all the reaction smiles that involve acetic acid (or its derivatives) from a table that stores reactions.

Reaction SMILES represents a specific reaction between specific reactants yielding specific products. As such, it can be very useful to store a library of reactions of interest. These might be a record of reactions that have been carried out at a company, a set of reaction plans in an academic research group, or even a set of hypothetical reactions that might never succeed in the laboratory. There is another extension of SMILES called SMIRKS that is more general and can represent a class of reactions.

SMIRKS mixes elements of SMARTS with SMILES. A more complete definition of SMIRKS is available at daylight.com.[1] Several authors discuss the graph theory underpinnings of SMIRKS.[2] The reaction SMILES above, CC(=O)O.CN>>CC(=O)NC.O, could be generalized to allow acid chlorides as well as carboxylic acids. This would be CC(=O)[O,Cl].CN>>CC(=O) NC.[*]. The use of the SMARTS [O,Cl] allows oxygen or chlorine. The second product, which was water in the initial example, is expressed simply as [*] since it could be either Cl or O. When SMIRKS is used, it cannot be considered a type of SMILES and therefore cannot be used in the matches function or other functions requiring SMILES. However, SMIRKS can be used to transform a given set of SMILES according to the rule specified in the SMIRKS: in other words, to carry out the transformation in silico.

9.3   Transformations

The word *reaction* is typically used to represent a specific reaction as well as a general transformation. In this chapter, the word *reaction* is used to mean a specific reaction represented using reaction SMILES. The word *transformation* is used to indicate a change in a set of reactants to products. SMIRKS is the language used to specify precisely how this transformation is to be carried out.

In order to carry out a transformation, it is necessary to know precisely which bonds are to be broken and which are to be made. While this information is implicit in the SMIRKS above and can be understood by any chemist, a more specific set of instructions is necessary in order to make the transformation possible using a computer. Numbering the atoms of the reactants and the corresponding atoms in the products accomplishes this. This produces an atom mapping. Many sketching programs can do this automatically or with some additional input from the user. Using the example shown previously, an atom-mapped SMIRKS would

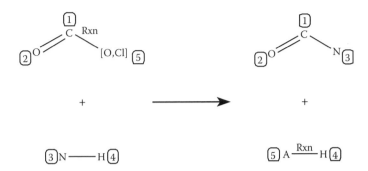

Figure 9.1 Atom-mapped reaction of an amine with an acid or acid chloride.

be [C:1]([O,Cl:5])=[O:2].[N:3][H:4]>>[N:3][C:1]=[O:2].[*:5][H:4]. Figure 9.1 shows a depiction of this SMIRKS reaction. Notice that the second product, which was water in the initial example, is now expressed as [*:5][H:4]. This [*:5] is necessary because the second product could be either water or HCl. The use of [H:4] completes the full accounting of every atom involved in the transformation. Hydrogen atoms not involved in the transformation need not be explicitly specified. This atom-mapped SMIRKS has also left out 2 extra carbon atoms, one attached to each reactant in the original example. These carbons properly belong to the specific reaction SMILES discussed above, but they do not participate in the transformation and need not be specified in the SMIRKS.

9.3.1 Unimolecular Transformations

Before considering how SMIRKS can be used to carry out transformations with multiple reactants, first consider simpler unimolecular transformations. These are discussed separately because of the important use of unimolecular transformations to enforce the consistent use of SMILES throughout the database. This improves the integrity of the data in a chemical sense, rather than a relational database sense as discussed previously. The root of the issue is this: There are multiple ways to represent the same molecular structure due to the limitations of valence bond theory. In valence bond theory, upon which SMILES is based, atoms have formal charges, most often zero. The bonds between atoms are shared pairs of electrons and may consist of multiple shared pairs giving rise to double, triple, or possibly even higher-order bonds between atoms. This simple theory, while quite powerful and applicable to a majority of chemical structures, leads to certain ambiguities.

It is generally conceded that simple valence bond theory cannot adequately explain the bonds between the carbon atoms in benzene. This classic conundrum is often resolved by stating that there is a sort of mixing, or resonance of states. These two states can be represented using SMILES as C1=CC=CC=C1 and C1C=CC=CC=1. SMILES resolves this issue by introducing an extension to simple valence bond theory, namely an aromatic bond. The atoms on either end of the bond are referred to as aromatic atoms and are represented using the lowercase atomic symbol. So benzene becomes c1ccccc1 with implied aromatic bonds between the aromatic atoms instead of implied single bonds between nonaromatic atoms. Canonical SMILES performs this aromatization as well as reordering the atoms into canonical order. A separate unimolecular transformation is not necessary.

A more difficult issue arises for atoms that have different valence states. For example, nitrogen is typically considered to have a valence state of 3. The valence state of an atom is defined as the sum of the bond orders to the atom, minus the formal charge. So, ammonia has three single bonds and the nitrogen has valence state 3. Hydrogen cyanide has a triple bond to the nitrogen, again resulting in a valence state of 3. A nitrogen atom with a single bond and a double bond also yields a valence state of 3. Finally, the ammonium cation has four single bonds, but with a +1 formal charge on the nitrogen, yielding a valence state of 3. In some cases, it is desirable to consider nitrogen to have a valence state of 5. One common example is the nitro group, an example of which is CN(=O)(=O) in nitromethane. In this representation, the nitrogen has a valence state of 5. However, one might also use the SMILES C[N+](=O)[O−], which shows nitrogen in the more common valence state 3. Which SMILES is better? Unfortunately, there is no generally agreed-upon answer. Some prefer the charge-separated form because it reflects the more common valence state of nitrogen. Others prefer the former SMILES because it does not introduce formal atomic charges. In a sense, the answer is unimportant and is just a theoretical argument. Yet in a real-world database, it is important to have a consistent representation of any unique molecular structure.

One way to resolve this issue in a database is to require one particular form for the nitro group. Putting the burden on the chemist who inputs the structures is possible, but when hundreds or thousands of structures need to be imported, say from a vendor or other library, examining and correcting hundreds of individual structures is not feasible. Using a SMIRKS transformation can easily solve this problem.

Suppose it is decided that the valence 5, noncharge-separated representation of the nitro group is to be used throughout the database. The SMIRKS [O:2]=[N+:1][O−:3]>>[O:2]=[N+0:1]=[O+0:3], when applied to any charge-separated nitro group will transform it into the proper form. This is accomplished by creating another new SQL function, xform(smiles, smarts). As with the cansmiles and matches functions, this is an extension to standard SQL. Some form of this transformation function is

Table 9.1 Example SMIRKS for SMILES Standardization

Name	SMIRKS
nitro	`[NX3+:1](=[O:2])-[O-;X1:3]>>[N+0:1]`
	`(=[O:2])=[O+0:3]`
oxo-enol	`[C:1]=[C:2][OH1:3]>>[H][C:1][C:2]=[Oh0:3]`
sulfuro	`[SX4+:1](=[O:2])-[O-;X1:3]>>[S+0:1]`
	`(=[O:2])=[O+0:3]`
azide	`[NX1H0:1]#[NX2H0:2]=[N:3]>>[N-:1]=[N+:2]=[N:3]`

available in the chemical cartridges or extensions from several software vendors. The `xform` function is not contained in the core functionality shown in the Appendix.

Using the nitro transformation, the following SQL

```
Select xform('C[N+](=O)[O-]', '[O:2]=[N+:1][O-:3]>>[O:2]=[N+0:1]=
   [O+0:3]');
```

returns the SMILES CN(=O)=O. This approach can be used for any number of transformations in order to standardize the SMILES in any table of the database. Aside from nitro groups, there are other common variations in SMILES due to valence bond variations. Table 9.1 presents several of these and the SMIRKS that can be used to standardize them to a common form. These are just examples and need to be considered carefully for any application. With a table such as this, the following SQL could standardize an entire table of SMILES.

```
Update rxntest Set smiles=xform(smiles,smirks) From std_smirks;
```

where `std _ smirks` is the name of the table. Of course, whenever updating an entire table, great care should be taken that there are no unintended side effects. For example, the oxo-enol transformation is not favored by some chemists and is actually a tautomerization and not a valence bond issue.

It might be useful to prevent nonstandard SMILES from ever being inserted into a table. One way to do this is by using an SQL constraint. For example, an attempt to insert nonstandard SMILES into the following table would cause an SQL error.

```
Create Table atable (smiles Text Check (is_std_smiles(smiles)));
```

Another approach is to always use a function to insert SMILES into the table. For example:

```
Insert Into atable Select make_std_smiles(smiles);
```

Examples of the is _ std _ smiles and make _ std _ smiles functions are not shown here because neither of these approaches is ideal. In
the first case, using a check constraint, the nonstandard SMILES would
not be inserted, but the user would still be responsible for standardiz-
ing the SMILES and attempting the insert again. The second case using
a function is better, but it would still be possible to accidentally insert a
SMILES directly without the make _ std _ smiles function.

A better way to ensure chemical integrity of the SMILES is to use the
SQL trigger mechanism. An SQL trigger allows a function to intercept
data before it is inserted or updated and modify it if necessary.

Consider the following SQL, which uses PostgreSQL syntax.

```
Create Table atable (smiles text, id integer);
Create Function standardize() Returns Trigger As $EOSQL$
 Declare
   std_smiles Text;
   smirks Text;
   std Record;
 Begin
   For std In Select * from std_smirks Loop
     std_smiles = xform(NEW.smiles, std.smirks);
     If std_smiles != NEW.smiles Then
      NEW.smiles = std_smiles;
     End If;
   End Loop;
   Return NEW;
 End;
$EOSQL$ Language plpgsql;

Create Trigger standardize Before Insert Or Update On atable
 For Each Row Execute Procedure standardize();
```

The standardize function checks whether any smirks in the std _
smirks table, when used in the xform function, results in a modification of the input SMILES stored in NEW.smiles. If the xform function
does return a transformed SMILES, then that transformed value is used
in place of the value the user attempted to insert. Finally, the trigger is
created using the standardized function to possibly modify any SMILE
before inserting or updating a table.

9.3.2   Multi-Component Transformations

SMIRKS allows one to express a multicomponent transformation as well
as unimolecular transformation as discussed previously. The following
SMIRKS shows how to transform a combination of an acid chloride and
an amine into an amide.

[C:1][C:2]([O,Cl:3])=[O:4].[C:5][N:6][H:99]>>[C:5][N:6][C:2]([C:1])=[O:4].[*:3][H:99]

The unimolecular form of the xform function will not properly work with this SMIRKS. Instead, an alternate form that requires an array of SMILES is used. The following SQL

```
Select xform(ARRAY['CC(=O)Cl','CNC'],
'[C:2]([O,Cl:3])=[O:4].[N:6][H:99]>>[N:6][C:2]=[O:4].[*:3][H:99]');
```

returns CC(=O)N(C)C.Cl. With this xform function, it is possible to easily create combinatorial libraries from tables of reactants in the database. The following SQL will produce 100 products by combining 10 secondary amines and 10 acid chlorides from the table nci.structures.

```
select amine.smiles, acid.smiles, xform
  (ARRAY[amine.smiles,acid.smiles],
  '[H:99][N:1].[C:2](=[O:4])[Cl:3]>>[N:1][C:2]=[O:4]') from
(select smiles from structure where
  matches(smiles,'C[NH1]') limit 10) amine,
(select smiles from structure where
  matches(smiles,'C(=O)Cl') limit 10)acid;
```

This SQL statement can be expanded in many different ways to satisfy many different requirements. For example, an additional where clause in the sub-select statements could limit selection of reactants by molecular weight, cost, availability, etc. The type of amine or acid chloride could also be selected by changing the SMARTS in the matches function. For example, aromatic amines could be selected by using matches(smiles, 'c[NH1]').

It is possible to create a table containing many different SMIRKS and select the appropriate one by name. For example:

```
select amine.smiles, acid.smiles,
  xform(ARRAY[amine.smiles,acid.smiles], smirks) from
(select smiles from structure where
  matches(smiles,'C[NH1]') limit 10) amine,
(select smiles from structure where
  matches(smiles,'C(=O)Cl') limit 10) acid,
smirks_lib where smirks_lib.name='Schotten-Baumann';
```

Many other uses of the xform function are possible. Because the function is an extension of SQL, it can be easily used with all the other features of the SQL language and capabilities of an RDBMS.

There is no limit on the number of reactants or products that can be used in SMIRKS. The Ugi reaction[3] is a four-component condensation reaction combining an aldehyde, amine, carboxylic acid, and an isocyanide. The SMIRKS representing this transformation is [*:1][C:10](=[O:11]).[*:2][N:20]([H:21])[H:22].[*:3][C:30](=[O:31])[O:32][H:33].[*:4][N+:40]#[C-:41]>>[*:3][C:30](=[O:31])[N:20]([*:2])[C:10]([*:1])[C+0:41](=[O:32])[N+0:40]([H])[*:4]. In previous examples, every reactant atom was accounted for in the product.

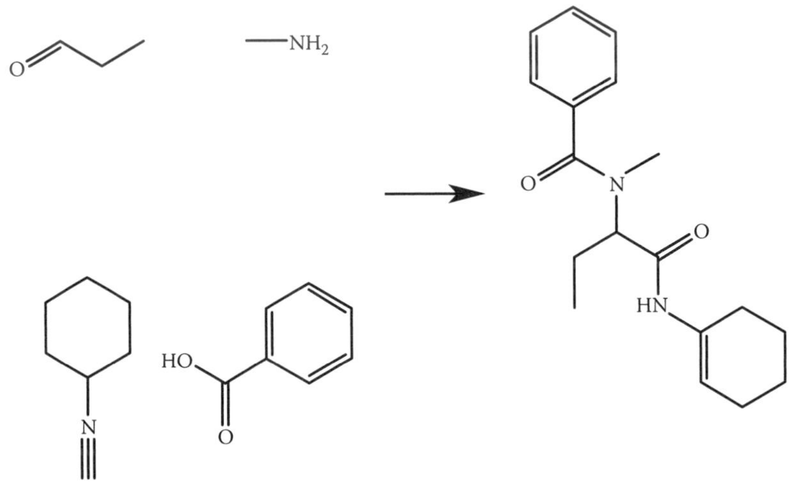

Figure 9.2 Four-component Ugi reaction.

In this case, reactant atoms that do not actually appear in the product are simply not mentioned in the product side of the SMIRKS. This is a useful feature of SMIRKS, allowing some reactant atoms to "disappear." This is convenient with hydrogen atoms, water oxygen atoms, and sometimes with other atoms. Depending on how the product SMILES will be further processed, it may be desirable to use this feature in order to retain only the main product. In addition, an unmapped hydrogen atom appears in the product on the [N:40] atom coming from the isocyanide. It is not important, and indeed probably not clear from the mechanism, exactly which hydrogen atom this is. For the purposes here, it is not crucial to exactly map that product hydrogen atom to a particular reactant hydrogen atom.

If the before-mentioned SMIRKS is stored in the smirks_lib table, then the following SQL

```
select xform(
 ARRAY['CCC=O','CN','c1ccccc1C(=O)O','[C-]#[N+]C1=CCCCC1'], smirks)
 from smirks_lib where name='Ugi4CC';
```

will return CCC(N(C)C(=O)c1ccccc1)C(=O)NC2=CCCCC2. This transformation can be drawn as shown in Figure 9.2.

9.4 *Canonical Reaction SMILES*

Reaction SMILES is a syntactical extension of regular SMILES. Since it is composed of regular SMILES punctuated with periods and the >>

symbol, each reactant and product can have a canonical representation. The overall reaction SMILES also has a canonical representation that is not necessarily the combination of the canonical SMILES of the reactants and products. The SQL extension function cansmiles correctly computes the canonical reaction SMILES. This can be used as a unique representation of a reaction in the same way that a regular SMILES can be used as a unique representation of a chemical structure.

Because SMIRKS is a combination of SMILES and SMARTS and because there is no canonical representation of SMARTS, there is no canonical representation of SMIRKS. SMARTS can be considered as a set of instructions on how to match substructures of SMILES. SMIRKS can similarly be considered as a set of instructions on how to identify reactive atoms and combine or alter them in order to carry out a specific transformation of a set of SMILES.

References

1. SMIRKS tutorial. 2007. http://www.daylight.com/dayhtml_tutorials/languages/smirks/index.html (accessed April 18, 2008).
2. Yadav M., Kelley, B.P. and Silverman, S.M. 2004. The potential of a chemical graph transformation system. In *Graph Transformations: ICGT 2004 Proceedings*, ed. H., Ehrig, G., Engels, F., Parisi-Presicce and G., Rozenberg, 83–95. Berlin: Springer Verlag.
3. Ugi, I., 1962. The α-addition of immonium ions and anions to isonitriles accompanied by secondary reactions. *Angewandte Chemie International Edition in English* 1(1):8–21.

chapter 10

PostgreSQL Extensions

10.1 Introduction

The basic capabilities of an RDBMS are accessible using the structured query language (SQL) language. These capabilities include the basic data types, such and `numeric`, `text`, `date`, etc. and basic functions and operators, such as `length`, `sqrt`, `=` and `like`. It is possible to extend the capabilities of the database and of SQL by defining new data types and new functions. These integrate neatly into the syntax of SQL and allow the new data types and functions to be easily used in ways similar to the standard SQL data types and functions. The PostgreSQL relational database management systems (RDBMS) allows the use of various computer languages to create new functions, including a procedural language plpgsql native to PostgreSQL. The plpgsql language is analogous, but substantially different than the sqlplus language used in the Oracle RDBMS. Of course, it is possible to simply use SQL to define new data types and functions as well. In this chapter, the focus is on PostgreSQL and the various languages available to extend its functionality.

Chapter 3 showed how SQL could be used to write a function to convert pressure data values expressed in atmospheres to kilopascals. Other functions were used in check constraints on a column containing CAS numbers. This chapter will show how new data types can be defined. This will require functions to define the method for input parsing and the method to output data values. There will also be functions to define operations on the new data types, enabling searches to be integrated easily with standard SQL syntax.

10.2 Composite Data Types

A composite type is defined in terms of existing data types. For example, the following SQL defines a new data type for concentration values.

```
Create Type conc As (val float, unit text);
```

This definition is similar to how a table is defined with its columns having names and data types. It might be possible to collect all concentration values into one table, but concentration values are very common and used in many different ways. They are likely to appear in many tables of a chemical database. Defining a new data type that can store both the concentration values and its units can be very helpful. It keeps the value and the units tightly coupled rather than stored in separate columns of a table.

Consider the following table creation for biological data.

```
Create Table assay1 (id integer, ki float, ki_unit text,
    ic50 float, ic50_unit text, ec50 float, ec50_unit text);
```

It is essential to keep the association of ki _ unit with ki in order to accurately express the value. It is also important that units for one data value are not accidentally associated with those for a different column. Naming the corresponding columns as above (ki and ki _ unit, ec50 and ec50 _ unit) helps, but using a composite data type actually enforces the correct association. This is another example of how database integrity can be increased. When the conc data type is used, this table becomes:

```
Create Table assay1 (id integer, ki conc, ic50 conc, ec50 conc);
```

The units of ki are kept associated with the ki values inside the conc data type.

When using a composite data type, the external representation of the value is different than the basic SQL data types. The components are represented as usual for number and text data types, but parentheses are used to associate the component values. For example, (1.74,nM) is the external representation of the conc value 1.74 nanoMolar. The following SQL produces sample output for an arbitrary compound id.

```
Select ic50, ec50 From assay1 Where id = 47665;
    ic50      |     ec50
-----------+-----------
(12.4,nM)   |   (1.33,uM)
(10.9,nM)   |   (2,uM)
(15.,nM)    |   (1.5,uM)
```

The individual components of the composite data type are also accessible using SQL. In this way, the output format can be altered and the individual components can be used anywhere in an SQL statement. For example, the following SQL produces sample output as shown below.

```
Select (ic50).val as "ic50(nM)", (ec50).val||(ec50). unit as ec50
From assay1 where (ic50).val < 10 and (ic50).unit = 'nM';
ic50(nM)    |     ec50
-----------+-----------
   8.5      |   1.23nM
   0.7      |   250uM
   1.4      |   87uM
```

10.3 Composite Data Type for Experimental Values

In most databases, a single value is used to represent an experimental measure. In many cases however, that value is meant to represent an upper limit or lower limit. For example, when measuring an IC50, the assay is sometimes limited in sensitivity and results are reported, for example as <0.15. If only the value 0.15 is stored, this would be indistinguishable from results where the value was measured to be equal to 0.15, or >0.15. This situation is typically handled by creating another column containing a symbol, < or > or null. When this is done, it complicates the search of these values because two columns must be considered and there is no built-in SQL function to perform the search required. This situation can be handled neatly by introducing a new SQL data type that incorporates a data value and a flag to denote < or > values.

The following SQL defines the type range, creates a sample table, and shows a selection of data from the table. Populating the table with data is not shown here.

```
Create Type range As (op Text, val Float);
Create Table rangetest (smiles text, ic50 range, name text);
Select * from rangetest;
smiles          |    ic50     |          name
----------------+-------------+--------------------------
BrC(Br)C(Br)Br  |   (=,10)    |  1,1,2,2-tetrabromoethane
OCCSCCCCCCCC    |   (<,10)    |  2-(octylthio)-ethanol
NCCCCCCCC       |   (>,10)    |  n-octylamine
```

The = is used to denote exact values. As with the conc values described above, keeping the value and the operator together in the same composite data type is preferred over keeping them in separate columns, especially when multiple value-operator pairs exist in the same table.

The external representation of this data type uses parentheses. This can be awkward, so the following input and output functions are defined.

```
Create Function range_parse(text) Returns range As $$
Select Case
  When substring($1, 1, 1) = '<' Then ('<',substring($1, 2))::range
  When substring($1, 1, 1) = '>' Then ('>',substring($1, 2))::range
  When substring($1, 1, 1) = '=' Then ('=',substring($1, 2))::range
  Else                                ('=',substring($1, 1))::range
  End;
$$ Language SQL;

Create Function range_text(range) Returns Text As $$
Select Case
  When ($1).q = '=' Then (($1).v)::Text
  Else              (($1).q||($1).v)::Text
  End;
$$ Language SQL;
```

These help in using the range data type, for example:

```
Insert Into rangetest (ic50) Values ( range_parse('<27.5') );
Select range_text(ic50) from rangetest;
 ic50   |
--------+
   10   |
  <10   |
  >10   |
 <27.5  |
```

It is always possible to use the ordinary external representation of the range data type using parentheses, but using range _ format and range _ text conforms to more common representations of data like these. It is possible to automate the range to text conversion even more, using the create cast SQL command, as follows.

```
Create Cast (range as text) With Function range_text(range)
   As Implicit;
```

When this is done, the range _ text function is implicitly called whenever necessary. So, the following SQL would produce the same table as shown above.

```
Select ic50::text from rangetest;
```

Since the range data type is not a standard SQL data type, the standard SQL operators cannot be used with this data type. However, new operators can be defined using SQL functions. Since the range data type is so similar to the float data type, implicit conversion to float is appropriate. This automatically allows many of the standard SQL operators to work with exact(=) range values.

```
Create Function range_float(range) Returns Float As $$
 Select Case When ($1).q = '=' Then ($1).v End;
$$ Language SQL;
Create Cast (range as float) With Function range_float(range)
   As Implicit;
```

As with the range _ text conversion, the range _ float conversion allows range values to be converted to float, whenever possible. This makes the following SQL work without explicit definition of the sqrt function for range data types.

```
Select ic50, ic50::text, ic50::float, sqrt(ic50) from rangetest;
ic50      |   ic50    |   ic50    |   sqrt
----------+-----------+-----------+--------------------
(=,10)    |    10     |    10     |  3.16227766016838
(<,10)    |   <10     |
(>,10)    |   >10     |
(>,27.5)  |  <27.5    |
```

Since the range _ float function only interprets exact(=) range values as float, null is returned whenever a range value contains < or >. So any function like sqrt, will also return a null value. Since range values can now be implicitly interpreted as float, many useful functions, such as sqrt, max, and avg and operators such as +, – and * become available for range data. These return the expected value for exact ranges, but null for < and > values.

The < and > operators of SQL can also be used with range data. However, since the conversion from range to float returns null for range data containing < and >, the following SQL only selects exact range values.

```
Select ic50::text, ic50::float, sqrt(ic50) from rangetest
  Where ic50 > 10;
```

If the table contains range values, such as 20 or 30, these will be correctly selected. However, if the table also contains range values such as >10, >20, and >30, these will not be selected using the above SQL.

In order to properly compare range values it is necessary to define functions that operate directly on the range data type, rather than indirectly after the implicit conversion to float. The following functions define how two range values should be compared for equality, less than, greater than, etc.

```
Create Function range_cmp(range, range) Returns Integer As $$
Select Case
 When ($1).q = ($2).q And ($1).v = ($2).v Then  0
 When ($1).q = '=' And ($2).q = '=' And ($1).v < ($2).v Then -1
 When ($1).q = '=' And ($2).q = '=' And ($1).v > ($2).v Then  1
 When ($1).v = ($2).v And (
  ( ($1).q = '<' And ($2).q = '=' ) Or
  ( ($1).q = '<' And ($2).q = '>' ) Or
  ( ($1).q = '=' And ($2).q = '>' )) Then -1
 When ($1).v = ($2).v And (
  ( ($1).q = '>' And ($2).q = '=' ) Or
  ( ($1).q = '>' And ($2).q = '<' ) Or
  ( ($1).q = '=' And ($2).q = '<' )) Then  1
 When ($1).v < ($2).v And (
  ( ($1).q = '<' And ($2).q = '=' ) Or
  ( ($1).q = '<' And ($2).q = '>' ) Or
  ( ($1).q = '=' And ($2).q = '>' )) Then -1
 When ($1).v > ($2).v And (
  ( ($1).q = '>' And ($2).q = '=' ) Or
  ( ($1).q = '=' And ($2).q = '<' ) Or
  ( ($1).q = '>' And ($2).q = '<' )) Then  1
 Else Null
 End;
$$ Language Sql;
Create Function range_eq(range, range) Returns Boolean As $$
 Select ($1).q = ($2).q And ($1).v = ($2).v;
$$ Language SQL;
Create Function range_ne(range, range) Returns Boolean As $$
 Select ($1).q != ($2).q Or ($1).v != ($2).v;
```

```
$$ Language SQL;
Create Function range_lt(range, range) Returns Boolean As $$
 Select range_cmp($1, $2) = -1
$$ Language SQL;
Create Function range_le(range, range) Returns Boolean As $$
 Select range_cmp($1, $2) != 1
$$ Language SQL;


Create Function range_gt(range, range) Returns Boolean As $$
 Select range_cmp($1, $2) = 1
$$ Language SQL;
Create Function range_ge(range, range) Returns Boolean As $$
 Select range_cmp($1, $2) != -1
$$ Language SQL;
```

These functions could be used in SQL, but it is even more convenient to
define SQL operators that use these functions.

```
Create Operator = (Leftarg = range, Rightarg = range,
 Procedure = range_eq, Commutator = =, Negator = !=);
Create Operator != (Leftarg = range, Rightarg = range,
 Procedure = range_ne, Commutator = !=, Negator = =);
Create Operator < (Leftarg = range, Rightarg = range,
 Procedure = range_lt, Commutator = >, Negator = >=);
Create Operator > (Leftarg = range, Rightarg = range,
 Procedure = range_gt, Commutator = <, Negator = <=);
Create Operator >= (Leftarg = range, Rightarg = range,
 Procedure = range_ge, Commutator = <=, Negator = <);
Create Operator <= (Leftarg = range, Rightarg = range,
 Procedure = range_le, Commutator = >=, Negator = >);
```

Then, the following SQL selects the correct data from the table.

```
Select ic50::text, ic50::float, sqrt(ic50) from rangetest
 Where ic50 > range_parse(10);
 ic50    |   ic50    |     sqrt
---------+-----------+------------------
 >10     |           |
 20      |    20     | 4.47213595499958
 >20     |           |
 30      |    30     | 5.47722557505166
 >30     |           |
 99.3    |    99.3   | 9.96493853468249
```

Notice the use of range _ parse(10). This is necessary to force the comparison to be done using two range values. If the clause Where ic50 > 10
were used, the SQL parser might choose to convert the range value ic50
to float, rather than convert the constant 10 to range.

10.4 Array Data Types for Two- and Three-Dimensional Coordinates

For every standard SQL data type available in PostgreSQL, there is a corresponding array data type. While it is possible to define a composite data type for coordinates, consider using the array data type. For example:

```
Create table ctest (smiles text, name text, coord float[][3]);
```

The column `coord` is a two-dimensional array of `float` values. The external representation of array data uses curly braces. For example:

```
Insert Into coordtest (smiles,name,coord) Values
('CN1C=C(C)C(=O)NC1=O','1-methylthymine', '{
{-0.5223,-1.2374,0.2579},
{-1.8677,-1.3917,0.1177},
{-2.7245,-0.3893,-0.2291},
{-2.2127,0.8622,-0.4679},
{-0.8652,1.0783,-0.3284},
{0.0270,0.0456,0.0406},
{-3.9604,-0.6494,-0.3168},
{0.1995,-2.2271,0.5751},
{1.4782,0.3157,0.1674},
{-3.0687,1.9631,-0.8370}}');
```

The first dimension refers to the atom number and the second dimension refers to the Cartesian coordinates of the atom. So, the following SQL would select the smiles and the coordinates of the first two atoms of 1-methylthymine.

```
Select smiles, coord[1:2] From coordtest Where name='1-methylthymine';
```

Arrays are indexed starting with 1, rather than with 0 as is done in some computer languages. Notice that there are no array upper bounds specified for the first dimension of coords in the table creation. This allows each row to have a different number of atoms. The array _ upper function can be used to return the actual dimension used in any array.

The array _ upper function requires two arguments: the name of the array and the index for which the upper limit is requested. The function call array _ upper(coord,1) would return the number of atoms. The function call array _ upper(coord,2) would return 3. Even though the second index upper limit was specified as 3 in the table creation above, this was done for clarity and because the array was intended to hold 3-D coordinates. PostgreSQL does not enforce this upper limit. In fact, it would be possible to insert two-dimensional coordinates into the coordtest table. However, it is not allowed to mix two- and three-dimensional coordinates within any one array. Once the first atoms coordinates are given the insert statement, each succeeding atom must have the same dimensionality of coordinates.

It may be desirable to keep 3-D and 2-D coordinates in separate tables. However, it is possible to mix them in the same table. The following SQL will insert 2-D coordinates for 1-methylthymine into coordtest as created above.

```
Insert Into coordtest (smiles,name,coord) Values
('CN1C=C(C)C(=O)NC1=O','1-methylthymine', '{
{-0.0000,-0.8250},
{0.7145,-0.4125},
{0.7145,0.4125},
{0.0000,0.8250},
{-0.7145,0.4125},
{-0.7145,-0.4125},
{1.4289,0.8250},
{-0.0000,-1.6500},
{-1.4289,-0.8250},
{0.0000,1.6500}}');
```

The `array_upper` function can be used to determine the actual dimensions of an array. The following SQL would select only the 2-D coordinates arrays from the coordtest table.

```
Select smiles, coord from coordtest where array_upper(coord,2) = 2;
```

Each row in the `coordtest` table represents a molecule. The `smiles` column is a string of atom symbols and bonds and the `coord` column is an array of atom coordinates. How is it possible to keep the ordering of atoms in the smiles string in sync with the ordering of atom coordinates in the coord array? When the coordinates are initially entered from the external source, they are likely to be in a common chemical file format. The program that converts from that file format to SMILES would have to output the atom coordinates in the same order as the atoms in the SMILES.

In order to keep the appropriate SMILES associated with the corresponding coordinates, consider using a new data type. For example:

```
Create Type mol (smiles Text, coord Float[][3]);
```

This type could be used in a table that might also contain a canonical smiles column, or even other variants of SMILES if desired.

```
Create Table moltest (amol mol, name Text);
```

When this table is initially populated, the `amol.smiles` and `amol.coord` would be taken from the conversion program, as would the name.

```
Insert Into moltest (amol,name) Values (
('CN1C=C(C)C(=O)NC1=O','{
{-0.5223,-1.2374,0.2579},
```

```
{-1.8677,-1.3917,0.1177},
{-2.7245,-0.3893,-0.2291},
{-2.2127,0.8622,-0.4679},
{-0.8652,1.0783,-0.3284},
{0.0270,0.0456,0.0406},
{-3.9604,-0.6494,-0.3168},
{0.1995,-2.2271,0.5751},
{1.4782,0.3157,0.1674},
{-3.0687,1.9631,-0.8370}}'),
'1-methylthymine');
```

There are many ways in which coordinates might be stored and used in a chemical relational database. These are considered more fully in Chapter 11.

10.5   Functions in Other Languages

The functions in this book have so far used only SQL. Most of these have used standard SQL although some have made use of PostgreSQL extensions to SQL. PostgreSQL also allows functions to be written in other languages. The plpgsql procedural language extends standard SQL with common programming language constructs such as variables, if–then–else constructs, for and while loops, and exception handling. The plperl and plpython procedural languages are also available, but offer no special advantages, except of course to those programmers proficient in those particular languages. Finally, functions can be written in the C language. Since PostgreSQL itself is written in C, these functions use the very same data structures and functions as the RDBMS itself. Therefore, functions written in C offer the greatest advantage for speed and efficiency.

10.5.1   Plpgsql

Creating a function in plpgsql is done in a way similar to the previous examples using the SQL language. The following function creation shows some of the useful features of plpgsql and differences from the SQL language.

```
Create Function center(xmol Mol) Returns Float[3] As $$
Declare
  centrum Float[3] :=  Array[0., 0., 0.];
  natoms Integer := array_upper((xmol).coord, 1);
  i Integer;
Begin
  For i In 1 .. natoms Loop
    centrum[1] = centrum[1] + (xmol).coord[i][1];
    centrum[2] = centrum[2] + (xmol).coord[i][2];
    centrum[3] = centrum[3] + (xmol).coord[i][3];

  End Loop;
  centrum[1] = centrum[1] / natoms;
```

```
  centrum[2] = centrum[2] / natoms;
  centrum[3] = centrum[3] / natoms;
  Return centrum;
End;
$$ Language plpgsql;
```

This function takes one argument of the composite data type mol. Notice that the argument(s) may be named in the function definition. Here, using (xmol mol) allows the variable xmol to be used as the name of the input argument instead of $1. The function returns an array of three floats computed as the center (average coordinate) of the input mol. There are two main sections to every plpgsql function. The Declare section contains variable names, their type, and initial values. The Begin section contains the code that performs the necessary operations. There should always be a Return statement. Using the moltest table defined above, the SQL statement

```
Select center(amol) From moltest Where name='1-methylthymine';
```

returns $\{-1.35168, -0.163, -0.10205\}$.

There are many other features of plpgsql that make it useful. There are variables that can hold entire rows of a table, loops that iterate over rows returned from a select statement, methods of handling errors, and ways of executing dynamically generated SQL strings. These are described in the on-line documentation[1] and books on PostgreSQL.[2]

10.5.2   *Plperl, Plpython, Pltcl*

There are several other procedural languages supported by PostgreSQL. Among these are perl, python, and tcl. None of these languages is inherently better than the others. It may be better to write functions using these languages for several reasons. One of the reasons for using a given language is a familiarity with the language. This might occur when a research group has been developing code in one particular language for years. A more compelling reason to use one language arises when there already exists a set of modules that compute the desired results. For example, there is a set of perl modules called PerlMol for molecular chemistry.[3] Consider how this can be used to write an extension functions for PostgreSQL.

Plperl is not actually different than perl. It simply defines a protocol for passing and returning arguments to and from a plperl function. This means that arguments passed to a plperl function use the PostgreSQL string external representation. For text and numeric arguments, this presents no problem, since perl routinely uses string representations, even for numerical data.

The following plperl function will compute a molecular formula from an input SMILES.

```
Create Or Replace Function MF(text) Returns Text As $EOPERL$
use Chemistry::File::SMILES;
use Chemistry::File::Formula;
my ($s) = @_;
my $mol = Chemistry::Mol->parse($s, format => 'smiles');
return $mol->print(format => formula);
$EOPERL$ Language plperlu;
```

Notice that the mechanism for passing arguments is identical to an ordinary perl function – using the @_ variable. The return value is simply a string.

Executing functions written in an interpreted language like perl or python is slower than using a compiled language. The `cansmiles` function written in plperl and using PerlMol (see Appendix) is many times slower than the `cansmiles` function available in CHORD from gNova, Inc. However, PerlMol is open source and contains many useful functions for processing molecular structures. On the other hand, the SMILES and SMARTS representations in PerlMol do not include stereochemical centers and have other limitations compared to the OpenEye OEChem library used in CHORD. As always, there are trade-offs involved in any decision about which computer languages and libraries to use for a particular project.

Another important consideration in selecting a procedural language for functions is the time it takes to develop the function. It is generally true that developing code in Perl or Python is faster than using a compiled language like C. It is also more complicated to handle input arguments and return values from C functions in PostgreSQL.

10.5.3 Core Chemical Functions

Chapter 7 introduced several core functions that serve as a foundation for a very useful chemical extension of an RDBMS. The CHORD cartridge from gNova, Inc. implements those functions and others. The previous section discussed a PerlMol implementation of some of the core functions. There are many other useful functions in the PerlMol modules. It is easy to build a very powerful extension of PostgreSQL using plperl and PerlMol. The Appendix shows how all of the functions described in Chapter 7 can be implemented using plperl and PerlMol. The Appendix also shows another open source implementation using plpythonu procedural language and the FROWNS toolkit.[4] The third implementation of the core functions shown in the Appendix uses plpythonu and OpenBabel.[5]

Table 10.1 summarizes the core functions that are implemented using PerlMol, FROWNS, and OpenBabel. The fingerprint `fp` function is not included in the core functions as defined in Chapter 7, but it is a very useful function for computing similarities of molecular structures and for speeding up structure matching using the matches function. The fingerprint function `fp` is not implemented using PerlMol. The `isosmiles`

Table 10.1 Core Extension Functions

Function Name	Returns	Arguments
valid	Boolean	smiles
cansmiles	Text	smiles
isosmiles	Text	smiles
keksmiles	Text	smiles
matches	Boolean	smiles, smarts
list_matches	Integer[]	smiles, smarts, imatch, istart
count_matches	Integer	smiles, smarts
molfile_to_smiles	Text	molfile text
smiles_to_molfile	Text	smiles text
fp	Bit	smiles, nbits, maxpath
contains	Boolean	bit, bit

function is not implemented using PerlMol or FROWNS. The `keksmiles` function is not implemented using FROWNS.

10.5.4   C Language Functions

It is possible to write PostgreSQL functions in C. Because PostgreSQL itself is written in C, any extension functions can take advantage of the internal representation of RDMBS data. In fact, examples on which C functions might be based are the very functions used by the PostgreSQL RDMBS for processing standard SQL data. This puts any C functions on an equal footing with any built-in SQL functions. For this reason, C functions are the best choice when issues of speed are important.

The CHORD[6] chemical cartridge is a commercial product from gNova, Inc. It is written using C functions and the OEChem toolkit from OpenEye. It provides the core functions discussed in this book, such as `cansmiles, matches, count _ matches, list _ matches, smiles _ to _ molfile, molfile _ to _ smiles`, and `xform`. CHORD makes it possible to efficiently process RDBMS tables containing many millions of chemical structures.

Unlike the procedural languages discussed above, C language functions are compiled separately. The code itself is not included in the SQL create function command. Instead, the `create function` command refers to a compiled object such as shared object (.so) file located in some directory on the server running the RDBMS. For example, the CHORD `oe _ smiles` function is defined as follows.

```
Create or Replace Function oe_smiles(Text, integer)
 Returns Text As 'gnova', 'oe_anysmiles'
 Language 'c' Immutable Strict;
Comment On Function oe_smiles(Text, integer)
 Is 'smiles to smiles of various types';
```

This defines the SQL function oe _ smiles, which is realized by calling the C function oe _ anysmiles in the shared object file gnova.so. The definition of the input arguments and return values is done in the same way as for a function written in any other procedural language.

There is an open-source extension of PostgreSQL called pgchem.[7] It uses C functions and OpenBabel. It implements some of the core functions described in Chapter 7. The names of the functions are not the same as the names used here. Using pgchem, it should be possible to perform most of the operations represented by the core functions described in this book.

The Appendix contains a simple C language function to return the number of bits set in a bit data type. The PostgreSQL Web site documentation contains examples of C language functions.

10.6 Object RDBMS

PostgreSQL is generally referred to as an Object Relational Database Management System (ORDBMS). The use of the word *object* implies objects in the sense of an object-oriented computer language. While not intended to be fully object-oriented in the same sense as a computer language, an ORDBMS shares the essential aspects of objects. These include composite data types, methods (functions), and inheritance.

This chapter has shown how composite data types can be of great use in chemical databases. The components of a composite data type are either basic SQL data types, or other composite data types. This is the same way an object is defined in an object-oriented computer language. This level of abstraction can help simplify the development of complex databases.

Another important aspect of objects is the methods that operate on them. An ORDBMS calls these functions, but the effect is the same. As shown in this chapter, the functions defined for new data types enable them to be integrated in the SQL language and handled just like standard SQL data types. This can be by casting the data type, for example, to allow range data to be treated as float. New functions that operate exclusively on range data can be defined, such as range _ cmp. The ability to define operators of new data types enhances their usability and integration into SQL even more.

An ORDBMS is a kind of hybrid of a traditional RDBMS and a fully object-oriented OODBMS. The range of types of DBMS and their advantages and disadvantages is discussed elsewhere.[8–10]

References

1. Plpgsql. 2008. http://www.postgresql.org/docs/8.2/static/plpgsql.html (accessed April 18, 2008).
2. Stones R. and Matthew, N. 2005. *Beginning databases with PostgreSQL.* Berkeley: Apress.

3. Tubert-Brohman, I. PerlMol—Perl Modules for Molecular Chemistry. 2006. http://www.perlmol.org/ (accessed April 18, 2008).
4. Kelley, B. FROWNS. 2002. http://frowns.sourceforge.net/ (accessed April 18, 2008).
5. Open Babel: The Open Source Chemistry Toolbox. 2008. http://openbabel. sourceforge.net/ (accessed April 18, 2008).
6. O'Donnell, T.J. CHORD. 2008. http://www.gnova.com/ (accessed April 18, 2008).
7. Schmid, E.-G. 2004. Pgchem::tigress extension to PostgreSQL. http:// pgfoundry.org/projects/pgchem/ (accessed April 18, 2008).
8. Grimes, S. 1998. Modeling object/relational databases. http://www.dbmsmag. com/9804d13.html (accessed April 18, 2008).
9. Object-relational database. 2008. http://en.wikipedia.org/wiki/Object-relational_database (accessed April 18, 2008).
10. Devarakonda, R. S. 2001. Object-relational database systems—The road ahead. http://www.acm.org/crossroads/xrds7-3/ordbms.html (accessed April 18, 2008).

chapter 11

Three-Dimensional Molecular Structure Tables

11.1 Introduction

In this chapter, various ways are discussed in which tables might be used to store three-dimensional molecular structures. In these tables, each row represents a structure. The columns contain molecular properties, which may consist of arrays of atom properties. In previous chapters, the use of new data types was introduced to improve the way some data are stored and searched. Array data types were suggested as a way of storing atomic coordinates for a molecule. In this chapter, other ways will be shown in which molecular structures can be stored and searched in a relational database management system (RDBMS). These include the use of simplified molecular input line entry sytem (SMILES) and entire files from external sources, such as molfiles or structure data files (SDFs). Methods are shown to input, output, convert, and search molecular structures from within the database. The structured query language (SQL) statements shown are valid in PostgreSQL but may not be valid using other RDBMS. For example, the array data type is implemented differently in Oracle compared with PostgreSQL.

11.2 Using Tables Instead of Files

Computer files are routinely used to store chemical information. It might seem that there is no practical alternative, since computer files are ubiquitous and deeply ingrained into our ways of thinking about computers and information. A file is an excellent way to temporarily store information in order to move it from one computer to another, for example, by e-mail, ftp, or http. Keeping a library of even just hundreds of files of molecular structures can be inefficient and confusing. Maintaining a schema of molecular structure tables in an RDBMS is efficient, structured, and reliable. Overcoming limitations of using many single files to store molecular structures have been attempted. For example, an SDF can store multiple molecular structures using a special symbol ($$$$) to separate structures within the file. It is also possible to store data about each structure by adding records to the file. But these additions to the file only increase its complexity and complicate (or even break) existing programs that need to read and write such files. For example, PDB files can be difficult to read and write, since there are many flavors of this "standard" file format with various additions to satisfy the needs of various computer programs. There are dozens of other molecular file formats, each with its own format.

One solution to the multitude of file formats for molecular structures is to provide a common program to read and write each file type.[1] A good example is Babel or OpenBabel.[2] A common data structure, internal to the program, serves as a hub for storing and processing the molecular structure. Components can be added to allow new file formats to be read and written. This approach shares some features with the RDBMS approach. Each molecular file format corresponds to an external representation of the molecular structure and the internal data structure corresponds to the internal representation. In the RDBMS approach, the various file formats are also the external representation of molecular structure, but the common data structure is a schema with tables holding the molecular structure information. The purpose of this chapter is to propose ways to move away from file formats entirely, preserving only the ability to read files formats for legacy data. A later section of this chapter will show how molecule tables in an RDBMS can effectively be used instead of molecular structure files by client programs.

11.3   Molfile and Other Common File Formats

The molfile or sdf file format is a very common way to store molecular structures. This can be considered as an external representation of a molecular structure data type. There are many other common file formats in use and only the essential features common to all of them will be considered here. The essential aspects of molecular structure contained in these files are atomic number or atomic symbol, formal atomic charge, bonded atom pairs, and bond orders. These are the minimum attributes necessary to define an unambiguous valence bond molecular structure. Other atom properties, such as atom types might also occur in these files, but these are specific to particular modeling programs and will not be discussed here. Sometimes molecular properties are also stored in these files. A way to store these properties in relational tables is discussed.

It would be possible to create tables using columns to store the atomic symbols and bond information found in molecular structure files, reflecting the column style format of the file itself. Instead, a SMILES representation of this valence bond information is preferred. SMILES is a compact text string containing the same information as the columns of atom symbols and bonds. It can also be used directly in the search functions described in earlier chapters. It is desirable to parse the molecular properties in molecular structure files in order to store them in data columns for possible searching using SQL. The properties are not stored as columns in the structure table. Instead, a separate table is created related to the structure table through the use of a structure id primary key. If necessary, the entire molecular structure file can be stored as a text string. This might serve as a repository of these files. It should be stored in a separate table related by use of a foreign key to a main structure table containing a unique primary key.

Although SMILES is an entirely equivalent way of storing a connection table of atoms and bonds, it is sometimes desirable to create a traditional connection table, for example, when an external program requires it. The extension functions `smiles _ to _ symbols` and `smiles _ to _ bonds` accept a SMILES string and produce an array of either symbols or bonds. These are discussed in a later section of this chapter. Several implementations of these functions are shown in the Appendix.

It may also be desirable to store the atomic coordinates read from these files. The purpose of parsing the coordinates from the file and putting them into a separate column is to enable use of the coordinates from within the database. If the column is properly defined as a `numeric` or `float` column, this will also ensure that the coordinates are proper numbers. If there is no need for atomic coordinates, it is not necessary to create a column for these. Later sections of this chapter will discuss ways in which these atomic coordinates might be used in SQL functions.

In a molecular structure file, an atom record typically contains all of the information about that atom: the atomic number or symbol, the charge, coordinates, etc. When such a file is parsed into a SMILES string and an array of coordinates, it is important to be able to associate the proper coordinate with the proper atom. The use of canonical SMILES ensures this. Because canonical SMILES defines a unique order of the atoms in a molecule, that order is used to store the coordinates. Later sections of this chapter will discuss ways in which atomic coordinates might be stored in columns of a table.

There are many programs available to parse the various molecular structure file format. OpenBabel is an open-source program that can read many file formats and produce a SMILES representation of molecular structure. There are many other commercial products that can do this as well. In the following examples, the OpenBabel/plpythonu implementation of molfile parsing will be used. This was introduced in Chapter 10. The code to define the necessary functions is shown in the Appendix.

11.4   Processing SDF Files

A common way of distributing structural and chemical data is in the form of an SDF file. An SDF file is a collection of compounds stored in molfile format and separated with a record containing the string $$$$. Many compound vendors make their libraries available this way. Many research publications include SDF files of structures and data. In the following

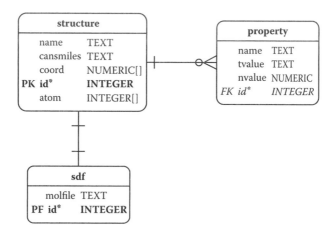

Figure 11.1 Entity relationship diagram for VLA4 schema.

example, SDF files were obtained from QSAR world,[3] a Web resource that curates dozens of data sets used in quantitative structure activity relationship (QSAR) studies. The VLA-4[4] Integrin antagonists were selected. This file contains structures and data for 94 compounds.[5]

One way to organize tables in a database is to define a new schema to contain related tables. Here, we will create a schema name vla4. Using an expansion of the example from the previous chapter, the following three tables are suggested as a starting point. The entity relationship diagram in Figure 11.1 illustrates the vla4 schema.

```
Create Schema vla4;
Create Table vla4.sdf (id Integer, molfile Text);
Create Table vla4.structure (id Integer, name Text, cansmiles Text,
  coord Float[][3], atom Integer[]);
Create Table vla4.property (id Integer, name Text, tvalue Text,
  nvalue Numeric);
```

The column structure.id is a unique integer relating the structure, sdf and property tables. The sdf.molfile column contains the molfile for each structure as defined by the vendor. The structure.name and structure.cansmiles columns contain the name and canonical smiles parsed and computed from the molfile. The structure.coord column will contain an array of atomic coordinates. The structure. atom column will contain an array of atom numbers from the file in canonical order to correspond to the atom order in the canonical SMILES. The OpenBabel/plpythonu extension functions molfile _ mol and molfile _ properties will be used to parse the vendor SDF molfiles and populate these tables. The molfile column of the sdf table is first populated from the SDF file, using the following perl script.

```
print <<EOSQL;
Create Schema vla4;
Create Sequence vla4.structure_id_seq;
Create Table vla4.sdf (id Integer
  Default Nextval('vla4.structure_id_seq'), molfile Text);
Create Table vla4.structure (id Integer Primary Key
  Default Nextval('vla4.structure_id_seq'), name Text, cansmiles
Text,
  coord Numeric[][3], atom Integer[]);
Create Table vla4.property (id Integer References vla4.structure (id),
 name Text, tvalue Text, nvalue Numeric);
Copy vla4.sdf (molfile) From Stdin;
EOSQL

while (<stdin>) {
 if (/\$\$\$\$/) {
  print;
 } else {
  s/\r//; chomp; print; print "\\n";
 }
}
```

The script contains a few SQL statements needed to create the schema
and tables. Notice that a named sequence, val4.structure _ id _ seq
is created. This is used to create a new structure.id whenever a new
row is added to either the vla4.sdf or vla4.structure table. The
structure.id column is chosen as the primary key. The SDF file is read
from standard input and separated into individual molfiles using the $$$$
delimiter. The output from this script, named loader, is piped into the
psql command as follows.

```
perl loader <vla-4.sdf | psql mydb
```

The vla4.structure table is chosen to contain the primary key instead
of the vla4.sdf table. This allows the vla4.sdf table to be dropped at a
later time without upsetting the relational integrity of the overall schema.
It would be possible to define the vla4.sdf.id column as a foreign key to
the vla4.structure.id column when the tables are created. However,
if that were done it would not be possible to insert into the vla4.sdf
file without a corresponding row in the vla4.structure table. After the
vla4.structure table is populated (see below), a foreign key constraint
will be added to the vla4.sdf table.

At this point, the vla4.sdf table has been created in the database
named mydb. The molfile column contains the molfile for each structure
in the sdf file. The sdf.id column contains a unique integer that can be
used to relate the vla4.property table. Notice the use of the default
value nextval('vla4.structure _ id _ sql') in the SQL statement
that creates the vla4.sdf table. This causes the sdf.id column to contain

a unique serial integer each time a row is inserted. The `structure.id` uses the same sequence, so that it may be kept in relational integrity with the `vla4.sdf` table. Eventually, the `sdf.id` column will become a primary foreign key having a one-to-one relationship with `structure.id`. This ensures that there must be exactly one sdf row for each structure row. The `property.id` column has a foreign key constraint related to the primary key column `vla4.structure.id`. This ensures relational integrity among the tables in this `vla4` schema.

The next step is to parse the molfile data into separate columns of the table. The `molfile_mol` function expects a molfile and returns a composite data type named `mol`. This data type is defined as:

```
Create Type openbabel.mol As (name Text, cansmiles Text, coords
    Float[][3], atoms Integer[]);
```

This composite type is created when the `openbabel` schema and its associated functions are created using the code contained in the Appendix. The atoms integer array is a map of the ordering of the atoms as they occur in the input file and the order as they occur in the canonical Smiles. This is not used here, but may be useful for other purposes. It is not necessary to issue the `create type` SQL command again during the loading of the data from the `vla-4.sdf` file. The data type may be used throughout the database, once it is created. It does not belong only to the schema `openbabel`, it simply resides there because its use is associated with functions in that schema. If this data type is used throughout the database, it might be moved to a more public schema.

The following SQL distributes the data returned from the `molfile_mol` function into the columns of `vla4.structure`.

```
Insert Into vla4.structure (id, name, cansmiles, coords, atoms)
Select id, (openbabel.molfile_mol(molfile)).* from vla4.sdf;
```

Notice that the individual elements of the composite data type value returned by the `openbabel.molfile_mol` function must be used. For example, `(openbabel.molfile_mol(molfile)).cansmiles` refers to the cansmiles element of the composite value returned by `molfile_mol` function. The function return value `(openbabel.molfile_mol(molfile)).*` refers collectively to all the elements of the composite data type. The function return value `openbabel.molfile_mol(molfile)` refers to the single composite data value itself. This cannot be used in the `insert` statement shown before, since the `insert` requires 5 data values: `id`, `name`, `cansmiles`, `coords`, and `atoms`. At this point, the table `vla4.structure` looks like the sample shown in Figure 11.2. Only the first 30 rows are shown and the data in each column is truncated for this preview.

id	name	coords	atoms
1	BMCL-1051-13	{{5.1482,-1.8181,0.0},{4.9347,-1.0212,0.0},{4.137...	{30,29,28,27,26,22,9,8,7,10,2,1,6,5,4,3,31,11,18,...
2	BMCL-1051-14	{{3.9243,-0.0108,0.0},{3.1274,0.2028,0.0},{2.6252...	{27,26,22,9,8,7,10,2,3,4,5,6,1,28,11,18,12,17,20,...
3	BMCL-1051-15	{{4.3368,-0.7252,0.0},{3.9243,-0.0108,0.0},{3.127...	{28,27,26,22,9,8,7,10,2,3,4,5,6,1,29,11,18,12,17,...
4	BMCL-1051-16	{{5.0512,-1.9627,0.0},{5.0512,-1.1377,0.0},{4.336...	{30,29,28,27,26,22,9,8,7,10,2,1,6,5,4,3,31,11,18,...
5	BMCL-1051-17	{{4.3368,-2.3752,0.0},{5.0512,-1.9627,0.0},{5.051...	{31,30,29,28,27,26,22,9,8,7,10,2,1,6,5,4,3,32,11,...
6	BMCL-1051-18	{{5.2656,-0.2363,0.0},{4.4476,-0.344,0.0},{3.9453...	{29,28,27,26,22,9,8,7,10,2,1,6,5,4,3,30,11,18,12,...
7	BMCL-1051-19	{{-1.0633,0.4729,0.0},{-0.2664,0.6864,0.0},{0.146...	{36,34,35,7,10,2,1,6,5,4,3,33,11,18,12,17,20,16,1...
8	BMCL-1051-20	{{-1.1944,0.6437,0.0},{-0.3975,0.8572,0.0},{0.185...	{37,35,36,7,10,2,1,6,5,4,3,34,11,18,12,17,20,16,1...
9	BMCL-1051-21	{{4.1379,-0.8076,0.0},{3.9243,-0.0108,0.0},{4.638...	{29,28,30,27,23,10,9,7,11,2,3,4,5,6,1,12,13,19,14...
10	BMCL-1051-22	{{-1.0633,1.6396,0.0},{-0.4799,1.0562,0.0},{-0.69...	{8,31,32,7,11,2,3,4,5,6,1,12,13,19,14,15,20,33,16...
11	BMCL-1051-23	{{4.8187,-2.7092,0.0},{5.4021,-3.2925,0.0},{5.188...	{8,31,32,7,11,2,1,6,5,4,3,12,13,19,14,18,21,17,16...
12	BMCL-1051-24	{{4.6025,-2.7091,0.0},{5.1859,-3.2925,0.0},{4.972...	{8,32,33,7,11,2,1,6,5,4,3,12,13,19,14,15,20,34,16...
13	BMCL-1051-25	{{5.2656,-0.2363,0.0},{4.4476,-0.344,0.0},{5.031...	{30,29,31,28,27,23,10,9,7,11,2,3,4,5,6,1,12,13,19...
14	BMCL-1051-26	{{5.2656,-0.2363,0.0},{4.4476,-0.344,0.0},{3.9454...	{30,29,28,31,27,23,10,9,7,11,2,3,4,5,6,1,12,13,19...
15	BMCL-1051-27	{{5.7678,-0.8908,0.0},{5.2656,-0.2363,0.0},{4.447...	{32,30,29,28,31,27,23,10,9,7,11,2,3,4,5,6,1,12,13...
16	BMCL-1051-28	{{-1.0633,1.6396,0.0},{-0.4799,1.0562,0.0},{-0.69...	{8,32,33,7,11,2,1,6,5,4,3,12,13,19,14,15,20,34,16...
17	BMCL-1051-29	{{-1.0633,1.6395,0.0},{-0.4799,1.0562,0.0},{-0.69...	{8,34,35,7,11,2,1,6,5,4,3,12,13,19,14,15,20,36,16...
18	BMCL-1051-3	{{5.0512,-1.1377,0.0},{4.3368,-0.7252,0.0},{3.924...	{29,28,27,26,22,9,8,7,10,2,1,6,5,4,3,30,11,18,12,...
19	BMCL-1051-30	{{-1.0633,1.6395,0.0},{-0.4799,1.0562,0.0},{-0.69...	{8,35,36,7,11,2,1,6,5,4,3,12,13,19,14,15,20,37,16...
20	BMCL-1051-31	{{-4.0669,1.6256,0.0},{-3.4418,1.1875,0.0},{-3.44...	{9,30,31,8,7,5,6,1,2,3,4,19,20,29,21,22,26,32,23,...

Figure 11.2 Sample output from phpPgAdmin showing how coordinates and atom indexes are stored as arrays.

Next, the foreign key constraint is added to the `vla4.sdf` table. This ensures relational integrity between the `vla4.sdf` table and the `vla4.structure` table. The following SQL is used.

```
Alter Table vla4.sdf Add Constraint sdf_id_fk Foreign Key (id)
  References vla4.structure (id);
```

Finally, any properties contained in the molfile will be stored in a separate table containing the text value copied from the file as well as a numeric value for the property, if that is appropriate for the property. There will be a one-to-many relationship between the structure and property table, allowing any number of properties to be stored for each structure. The function `openbabel.molfile _ properties` is shown in the Appendix. It expects a molfile and returns a composite data type, defined as follows.

```
Create Type openbabel.named_property As (name Text, value Text);
```

This composite type is created when the `openbabel` schema and its associated functions are created using the code contained in the Appendix. It is not necessary to issue this SQL command again during the processing of the data from the vla-4.sdf file.

The following SQL is used to parse the `molfile` column from the `vla4.structure` table using the `openbabel.molfile _ properties` function. The `name` and `value` fields of the composite data type are inserted into the `vla4.property` table, along with the appropriate id selected from the `vla4.structure` table along with the `molfile` column.

```
Insert Into vla4.property (id, name, tvalue)
  Select id, (molfile_properties).name, (molfile_properties).value From
  (Select id, openbabel.molfile_properties(molfile) From vla4.sdf) atmp
```

This `insert` statement is a bit more complex than the one that inserted rows into the `vla4.structure` table. In that table, only one row was returned from the `molfile _ mol` function. The `molfile _ properties` function returns multiple rows for each molfile, when there are multiple properties for each molfile. The second `select` statement above (the one in parentheses and identified with the name `atmp`) selects all the rows for each molfile. The first `select` statement selects all the columns from each returned row from `molfile _ properties`. These are then inserted into the `vla4.property` table.

Finally, the nvalue column of the vla4.property table can be populated when possible. This column stores the numerical value of the property. Since not all values are numerical, this column may have null entries. The purpose is to enable efficient use of numerical data when appropriate, for example, to select by value, sort, apply mathematical functions, etc. The following SQL will update the nvalue column when possible with a numeric value. The tilde operator in the where selects text values that match the regular expression. The expression shown here allows integers, decimal values, and scientific notation using E or e for the exponent, for example 6.023E23.

```
Update vla4.property Set nvalue = tvalue::numeric
  Where tvalue ~ E'^[+-]?[0-9]+(\\\\.[0-9]*)?([Ee][+-]?[0-9]+)?\$';
```

It is possible to create additional columns in the property table, for example to contain an integer representation of the data value if that is considered necessary.

Once the data from the molfiles are extracted and loaded into the val4.sdf table, the table could be deleted. The `vla4.sdf` table might also be retained as a backup of the original sdf file, easily accessible from within the database for further processing. In either case, the `vla4.sdf` table will not create any overhead while using the `vla4.structure` or `vla4.property` tables. It will merely use space in the `vla4` schema, in the same way that a backup copy of the original sdf file would use space in a folder on the computer's disk.

It may seem that some information will have been lost if the original molfile is discarded. For example, the list of atomic symbols and bonds is not stored directly in the `vla4.structure` table. The cansmiles string, however, does contain this information. It may be necessary for some purposes to extract this information, for example, when an external program does not read SMILES but instead requires a list of atomic symbols and bonds. This is discussed more in the next section. A connection table can be generated from SMILES using the functions `smiles _ to _ symbols` and `smiles _ to _ bonds`. These functions are shown in the Appendix using both the FROWNS and OpenBabel toolkits. The following SQL produces a connection table in the form of an array of atom symbols and bond orders and indices.

```
select smiles,smiles_to_symbols(smiles),  smiles_to_bonds(smiles)
  from nci.structure where cas = '1467-70-5';
smiles                      |  smiles_to_symbols  |  smiles_to_bonds
--------------------------+---------------------+---------------------------
c1cc(oc1)C(=O)C(=O)O  |  {C,C,C,O,C,C,O,C,O,O}  |  {{1,2,4},{2,3,4},{3,4,4},…}
```

The `smiles _ to _ symbols` and `smile _ to _ bonds` functions return arrays of values. In the sample output above, the `smiles _ to _ bonds` output has been truncated for easier viewing. Some client programs may expect this information as separate rows, as if they were records in a file. These arrays may be cast into that form by using a plpgsql function that returns elements of an array as rows. This is shown in the next section.

11.5 Using Tables Instead of Files in Client Programs

If an RDBMS is used to store molecular structures, this change requires modifications to existing computer programs that read molecular structure files. The modifications are confined simply to the portions of the computer programs that read and write files. These portions become functions that use SQL to access an RDBMS. Chapter 5 introduced methods for client programs to access data stored in RDBMS tables using SQL. This section shows how an existing program that reads and writes molfiles can be readily modified to use an RDBMS. First, however, it is necessary to describe the schema and tables used to store molecular structures. It is important that these tables can accommodate not only information from molfiles, but also information from other molecular file formats in common use.

Consider the `vla4` schema described above. It might be possible for the client program to read the molfile data directly from the vla4.sdf file, but the goal is to use the data in the `vla4.structure` and `vla4.property` tables. Recall that these tables, or ones like them in another schema, could have been created from files other than molfiles. These tables could also have been populated with other client programs that no longer use files at all, but instead store molecular structure data in RDBMS tables.

A traditional client program reads from a molecular structure file and performs some computation that depends on the molecular structural data. This read(file) function reads particular columns or fields from the file. A different function would be necessary for each type of file format. A traditional client program can be modified to read molecular structure data from an RDBMS table by substituting another function for the read function. The getdata(name, table) function would select data from an RDBMS table. The following python code snippet shows one example of such a function.

```
def getdata(self, name, table):
    _insql = "select smiles_to_symbols(cansmiles) as symbols, coords
        from %s where name=''%s''" % (table, name)
    _sql = "select * from symbol_coords('%s') as (symbol text, x
        numeric, y numeri c, z numeric)" % (_insql)
    for (_sym, _x, _y, _z) in ((self.conn).query(_sql).getresult()):
        (self.symbols).append(_sym)
        (self.coords).append( (_x, _y, _z) )
    self.natoms = len(self.symbols)
    return
```

This method is part of a class containing self.natoms, self.symbols, and self.coords. The final piece to be explained here is the plpgsql function symbol _ coords. This function, shown in the Appendix, accepts an SQL statement that selects an array of symbols and an array of coordinates. These are then returned as rows in order to "read" them as if they were records in a file. The following command shows the first few rows output from the symbol _ coords function called using the command line psql function in a way similar to the getdata method above.

```
select * from
  symbol_coords('select openbabel.smiles_to_symbols(cansmiles) as
    symbols, coords from vla4.structure where name=''BMCL-1051-38''')
    as (symbol text, x numeric, y numeric, z numeric);
```

symbol	x	y	z
O	-4.067	1.6258	0.0
C	-3.4418	1.1875	0.0
O	-3.4418	0.3625	0.0
C	-2.7273	1.6	0.0
C	-2.7273	2.425	0.0
C	-3.4418	2.8375	0.0

Once methods like these are in place, they readily replace traditional read statements. RDBMS access methods are available for almost every programming language, not just python as in this example. These were discussed in Chapter 5.

11.6 *File Import, Export, and Conversions*

The purpose of this book is not to provide another method for interconversion of molecular file formats. The focus of this chapter it to provide enough background information to allow the database designer to create a schema of tables that can effectively represent molecular structure without regard to its external file format representation. It is often necessary to import a molecular structure stored in some particular file format. The extension functions `molfile _ to _ smiles`, `mofile _ mol`, and `molfile _ properties` accomplish this for molfiles. It is sometimes desirable to export a molecular structure using some particular molecular file format. The core extension function `smiles _ to _ molfile` will create a string representation of a molfile. The implementations of this function shown in the Appendix do not make use of the coordinates. It would be possible to create an import and export function for each molecular file format of interest. Using the OpenBabel implementation of the core extension functions would be an excellent starting point for such an exercise.

11.7 Functions Using Three-Dimensional Atomic Coordinates

The previous section shows how molecular structures stored in an RDBMS can be made available to client programs that traditionally read molecular structure files. The advantage of storing molecular structures in an RDBMS is that the information can be used from within the database, as well as by external clients. For example, it would be possible to search a table of molecular structures for three-dimensional overlap, much like it might be searched for substructure match. Of course, such search functions need to be written and installed as extensions to an RDBMS, just like the matches functions was done for substructure searches. This section shows some possible ways this might be accomplished.

There are many methods to overlap one molecule's three-dimensional coordinates onto another molecule's. Perhaps the simplest method simply finds the center of one molecule and translates the other molecule to that central coordinate. The following functions can be used to do this. The `align` function takes two arrays of coordinates and returns the difference between the two centers. This difference can be applied to either molecule to align it with the other. The functions `center` and `difference` are utility functions.

```
Create Or Replace Function align(amol float[][3], float[][3])
Returns float[3] A s $$
 Select difference(center($1), center($2));
$$ Language SQL Immutable;

Create Or Replace Function center(amol float[][3]) Returns float[3]
As $$
Declare
  centrum Float[3] :=  Array[0., 0., 0.];
  natoms Integer := array_upper(amol, 1);
  i Integer;
```

```
Begin
  For i In 1 .. natoms Loop
    centrum[1] = centrum[1] + amol[i][1];
    centrum[2] = centrum[2] + amol[i][2];
    centrum[3] = centrum[3] + amol[i][3];

  End Loop;
  centrum[1] = centrum[1] / natoms;
  centrum[2] = centrum[2] / natoms;
  centrum[3] = centrum[3] / natoms;
  Return centrum;
End;
$$ Language plpgsql Immutable;


Create Or Replace Function difference(float[3], float[3]) Returns
float[3] As $$
 Select ARRAY[$2[1] - $1[1],
              $2[2] - $1[2],
              $2[3] - $1[3]];
$$ Language SQL Immutable;
```

The `align` function might be used as follows, to align all structures in a table to a reference structure selected by name.

```
Select ref.name as ref, others.name, align(ref.coords, others.coords) from
(Select name, coords from vla4.structure where name = 'BMCL-805-1') ref,
(Select name, coords from vla4.structure) others;
```

The `align` function can be expanded in many ways. For example, instead of simply finding the center of each molecule, a substructure could be used. This might be defined as a SMARTS match that is expected in each of the molecules to be aligned. This would be a natural outcome of a substructure search. In order to create an array of coordinates for a subset of a molecule, the following function could be used.

```
Select ref.name as ref, others.name, align(ref.coords, others.coords) from
(Select name, subset(cansmiles,'CSC', coords) as coords
 from vla4.structure where name = 'BMCL-805-1') ref,
(Select name, subset(cansmiles, 'CSC', coords) as coords
 from vla4.structure Where matches(cansmiles, 'CSC')) others
```

This SQL statement is similar to the previous SQL statement, except that the subset function is used to select only those elements of the `coords` array for atoms that match the target of interest, here the substructure CSC. The `subset` function is defined as follows.

```
Create Or Replace Function subset(smiles text, smarts text, coords
float[][3]) Returns float[][3] As $$
Declare
 scoords Float[][3];
```

```
atoms    Integer[];
i        Integer;
j        Integer;
nmatch   Integer;
Begin
 Select gnova.list_matches($1, $2, 1, 1) into atoms;
 if atoms Is Null Then
  Return Null;
 End If;
 nmatch = array_upper(atoms, 1);
 i = atoms[1];
 scoords = coords[i:i][1:3];
 For j In 2 .. nmatch Loop
   i = atoms[j];
   scoords = scoords || coords[i:i][1:3];
 End Loop;
 --Return scoords[2:nmatch+1];
 Return scoords;
End;
$$ Language plpgsql Strict Immutable;
Comment On Function subset(text, text, float[]) Is
'Return subset of molecule''s atomic coordinates for atoms matching
SMARTS';
```

There are a great many more elaborate methods that might be used to align three-dimensional molecular structures. Each of these could be implemented as new SQL functions and used in SQL statements like the ones above to produce the alignment to be applied to each structure of interest.

11.8 Conformations

When a project involves three-dimensional structures, it often includes multiple conformations for any particular structure. In previous sections of this chapter, an array of three-dimensional coordinates was stored for each structure. When a project needs multiple conformations, another table is needed to accommodate this. Instead of a coords column in a structure table, a conformations table will be used. The unique cid column in the structure table will function as a primary key related to a cid foreign key column in the conformation table. The entity relationship diagram in Figure 11.3 illustrates this. This effectively allows many conformations for any structure. Each one has an energy and comment

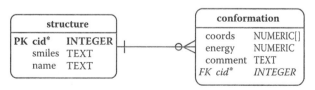

Figure 11.3 Entity relationship diagram for structures and conformations.

associated with it, as well as an array of coordinates. If more information is needed for conformations, additional columns can be added. Enough information must be stored in the conformation table to allow a meaningful selection. This might include a `method`, `date`, or any other identifying information.

In order to select conformations using the `vla4` schema described earlier, the following SQL might be used.

```
Select structure.cansmiles, structure.name,
 conformation.energy, conformation.coords
 From vla4.structure Join vla4.conformation Using (id)
 Where name = 'BMCL-1051-14';
```

This applies the same technique used throughout this book to `join` two related tables to select data from either table. This statement will select all the conformations for the named compound. Of course, further selection criteria could be added as desired to select the required conformation(s).

11.9   Other Representations of Three-Dimensional Molecular Structure

While atomic coordinates form the fundamental structure of a molecule, many methods prefer to represent a three-dimensional structure as a surface or a shape. Of course, these are ultimately computed from the atomic coordinates and perhaps atomic partial charges. It may be possible to represent these molecular surfaces or shapes as an array of three-dimensional coordinates. These could be stored as a column in the database analogous to the array of atomic coordinates. It might be necessary to create another data type, perhaps a composite data type, to store molecular surfaces or shapes. Once these representations are stored, they can be used in new SQL functions to assist in searching based on molecular surface or shape.

References

1. O'Donnell, T.J., Rao, S.N., Koehler, K., Martin, Y.C., and Eccles, B. 1990. A general approach for atom-type assignment and the interconversion of molecular structure files. *J. Comp. Chem.* 12(2):209–214.
2. OpenBabel. http://openbabel.sourceforge.net/ (accessed April 18, 2008).
3. QSAR World. http://www.qsarworld.com/ (accessed April 18, 2008).
4. Porter, J.R., Archibald, S.C., Brown, J.A. et. al. 2003. Dehydrophenylalanine derivatives as VLA-4 integrin antagonists. *Bioorg. Med. Chem. Lett.* 13(5): 805–808.
5. VLA4 dataset. http://www.qsarworld.com/qsar-datasets-porter.php (accessed April 18, 2008).

chapter 12

More on Client and Web Interfaces to RDBMS

12.1   Introduction

Most of the emphasis in this book has been on ways of using structured query language (SQL) and relational database management systems (RDBMS). Using functions written in SQL and other languages greatly extends the capabilities of existing RDBMS for handling molecular structures. Because the RDBMS is run as a server, a client program is necessary to interact with the RDBMS. Chapter 5 introduced several common client programs to do this. When developing a more complex system, existing applications may not satisfy the needs of the project. In that case, it becomes necessary to develop new client programs to interact with the RDBMS.

This chapter discusses ways in which more complex client applications can be written. These programs use SQL to select, insert, delete, or update tables in the database. Depending on the computer language used for the client program, a variety of interface libraries is available.

One advantage of using client programs is that they are independent of the RDBMS. For example, a client program written in Perl and using Perl::DBI could run equally well using an Oracle or PostgreSQL RDBMS to store the tables. Of course, there are some differences in SQL syntax among the various RDBMS. It is also possible to use certain unique features of one RDBMS that are not available in others. However, with some care, it is quite possible to write client programs that can easily run correctly when interfaced with most any RDBMS. Another advantage of using client programs is that they can be run on any client. This frees the database server to spend more of its resources on the database itself.

One disadvantage of using client programs is that data must be transferred to and from the server. Depending on how much data is required, this can cause a client program to run less efficiently than a server function run as an extension of the RDBMS.

Most new client programs will benefit from a judicious use of both new server-side SQL functions and new client functions. It is wise to carefully consider which operations are best done on the RDBMS server and which are done using a client program. There are several suggestions to consider in designing the best system for a project.

The following suggestions may help design and implement a better system for a project needing a client interface.

- Try to use SQL as much as possible.
- Consider extending SQL with new functions.
- Do not implement the same functionality in different clients.
- Only transfer data the user needs to see.
- Make good use of existing libraries.

Try to use SQL as much as possible. There are many useful features of SQL that can do much of what is needed when handling sets of data. For example, it would be wasteful to have a client program request an entire table in order to select certain rows. The SQL `select` statement is designed to efficiently search rows of a table. SQL can also sort selected rows using the `order` clause. When it is necessary to combine results from different queries, consider using the set operations of SQL, such as `intersect` and `union` rather than combining selected sets in the client program. There are other less obvious ways of using SQL to compute results. For example, the method used to compute polar surface area described in Chapter 8 makes use of SQL's ability to `join` tables and `sum` data in chosen columns. Not every computational chemistry method is amenable to use with SQL. For example, it is unlikely that a quantum mechanical energy could be efficiently computed using SQL compared to using a client.*

Consider extending SQL with new functions. This might be considered the fundamental suggestion in this book. There are many useful functions built into SQL, but sometimes a simple extension function can allow an SQL operation to run completely on the database server without having to pass data to the client. For example, to sort selected rows by and value in a column requires only simple SQL. If the data needed to sort the rows is not part of data being selected, consider writing a function that will provide the value to be sorted. For example, if it were necessary to sort by the number of atoms in a molecule, a `natoms(smiles)` function could be used in the `order` clause of SQL.

Do not implement the same functionality in different clients. If many different clients require the same functionality, it is better to encapsulate that in one central location—namely the RDBMS. For example, if it is necessary to compute molecular weight, it is better to have a server-side function do this in a consistent way rather than to implement such a function in each of the languages used for client applications. This may be more obvious for functions like fingerprints that are more difficult to re-implement in various languages. Moreover, it is essential that fingerprints be

---

* Since SQL `join` bears many similarities to matrix multiplication, this is not as crazy an idea as it may sound.

computed identically for identical molecules. Putting that functionality into the RDBMS itself and making it accessible through SQL is the best way to do this.

Only transfer data the user needs to see. There is always a need to transfer some data to and from the server. When a user enters data from a file or other source, it must be transferred to the database server in order to be inserted into the database. Similarly, when a user needs to see data values or a structural depiction, it must be transferred to the client application. However, if the data only needs to be processed by the client and then transferred back to the server, be sure to transfer only required columns. Also consider again writing a server function to process the data as required.

Make good use of existing client programs and libraries. There are many client programs that may already do what is required. Some examples of these were given in Chapter 5. There are also many client programs or libraries that may help in developing client applications using SQL to access the RDBMS. A discussion of these libraries for various languages follows. There is no inherent advantage to using any particular computer language to write a client program. If a particular project already has parts written in one language, it is probably best to continue new code development in that language. But if one language has features desirable for the application, that language may be the best choice. For example, many people prefer php for writing Web client applications that are run using CGI. If more interaction is necessary in the Web application, consider using a java applet. When the crucial parts (the data tables and functions) of the system are stored in a central location (the RDBMS), it becomes less important which language is used for client applications.

12.2   Store All Possible Data in the RDBMS

Before looking at specific examples of client programs, consider how and where data are typically stored and processed. Most computer data are stored in files. These are sometimes called flat files. This implies that there is no imposed structure in the files. The programs accessing the data infer any structure. On the other hand, data stored in an RDBMS table is highly structured. This requires some thought before the data is inserted into the table, but the benefit of structuring the data is great. When new data is received, it is likely to be in the form of a file, perhaps an e-mail attachment, a spreadsheet file, a CD, or download from a Web site. It is tempting (and common) to store these file on the computer and to process data in these files directly. It is much more advantageous if the data is put into an RDBMS as soon as possible. Consider the following scenario.

A colleague or coworker has just sent a file of molecular structures. They may need some feedback about which structures are desirable for a particular purpose, say, for possible purchase or for screening in some assay. Suppose the data are in the form of a molfile. If the structures and data in the file are added to an RDBMS having chemical extensions, the data become immediately more useful. The act of importing the molfile, say, to a table containing smiles, name, id, and columns of molecular data will immediately ensure several things. First, it will ensure that the molecular structure is valid and can be represented as simplified molecular input line entry system (SMILES). It will ensure that the data values required to be numeric truly are numeric. If there are chemical data types implemented in the RDBMS, constraints using those data types can be applied. The more chemical functionality there is in the RDBMS, the more information about the structures will become validated and easily available.

There are other advantages to importing structural data into the RDBMS as soon as possible. Depending on what other tables are in the RDBMS, it will now be easy to discover which structures are already contained in other tables of the RDBMS. The data in the new table will be easily accessible to other users and client applications. Once a decision has been made about which new structures are of interest, these can be readily moved to other tables in the RDBMS for further work (purchasing, testing, synthesis, etc.).

Another advantage of using an RDBMS to store chemical data is simply one of organization. It is very common to have dozens of files of molecular structures. One typically tries to remind oneself where the file came from, when it was received, what the purpose of the file is, etc. Using encoded names for the files or the folders containing the files is a typical approach. This quickly becomes unwieldy and confusing. On the other hand, if RDBMS tables are created to contain these data, sensible column and table names can be created to store information otherwise encoded in file and folder names. In addition, the generous use of table and column comments helps make sense of large amounts of data.

In short, it is possible and desirable to use an RDBMS to replace many, if not all, of the ways in which computer files are used. There are many advantages to using an RDBMS to store chemical data compared with using flat files. In spite of the familiarity of using file operations (open, read, write) in most client programs, these are easily replaced by SQL operations on the data in the RDBMS. In fact, many operations typically carried out by client programs can be done entirely within the RDBMS using chemical extensions and procedural languages to write new SQL functions.

12.3   Advanced SQL Techniques

Chapter 5 introduced ways in which client programs can be used with an RDBMS server. Some existing client programs, such as Excel and R, have methods to access data on an RDBMS server. Some Web applications, such as phpPgAdmin are specifically designed to interact with an RDBMS. When a new client program is needed, most computer languages offer a library that allows interaction with a server RDBMS using SQL. Chapter 5 introduced some simple examples in several languages. In those examples, a constant SQL string was used. In an actual client program, the SQL necessary to access the RDBMS will often need to be different, depending on input from the user. One obvious way to accomplish this is to use standard string operations to build a string from various SQL fragments, substituting the user-supplied values where necessary. This is entirely appropriate in many situations, but there are more efficient ways to use SQL and the various SQL modules and packages.

12.3.1   Placeholders in SQL Statements

Consider the following example using Perl to insert data into a table.

```
my $dbh = DBI->connect("dbi:Pg:dbname=$dbname;host=$host", $username,
$password);
while (($id, $ic50, $ed50) = &get_data()) {
  my $sql = "Insert Into test_assay (id, ic50, ed50)
            Values ($id, $ic50, $ed50)";
  my $sth = $dbh->prepare($sql);
  my $rv = $sth->execute;
}
```

The get _ data function is not detailed here, but would return three values, perhaps from user input, a file of data, or an instrument. The detail to notice in this example is that the call to $dbh->prepare is made once for every set of data values. The DBI prepare function is relatively inefficient. There is a more efficient way to insert multiple rows. The following code uses placeholders in the SQL statement.

```
my $sql = "Insert Into test_assay (id, ic50, ed50) Values (?,?,?)";
my $sth = $dbh->prepare($sql);
while (($id, $ic50, $ed50) = &get_data()) {
  last if ($id < 1);
  my $rv = $sth->execute($id, $ic50, $ed50);
}
```

In this example, the prepare function is executed only once. The SQL statement passed to the prepare function contains placeholders to represent values that will be made available once the statement is actually executed. The placeholder is simply a question mark. The arguments to the execute function provide three new values during each execution of the loop. This runs faster than the previous example.

The technique of using placeholders in prepared SQL statements is common in java as well. The following code snippets show examples in java.

```
PreparedStatement st = con.prepareStatement("Insert Into test_assay
(id,ic50,ed50) Values (?,?,?)");
while ( true ) {
  data_values = get_data();
  if ( data_values[0] < 1 ) break;
  st.setInt    (1, (int) data_values[0]);
  st.setDouble(2,       data_values[1]);
  st.setDouble(3,       data_values[2]);
  st.executeUpdate();
}
```

In this code snippet, the connection would already have been established and stored in the variable con. The con.prepareStatement method recognizes the use of the placeholder. The st.setInt and st.setDouble methods are more specific than in the perl example, requiring the integer or double data type.

The final language example shows how php supports SQL placeholders.

```
$sql = "Insert Into test_assay (id,ic50,ed50) Values ($1,$2,$3)";
$stmt = pg_prepare($dbconn, "test_assay_insert", $sql);
while ( $data_values = get_data() ) {
  if ( $data_value[0] < 1 ) break;
  $rv = pg_execute($dbconn, "test_assay_insert", $data_values);
}
```

The placeholders here are $1, $2, and $3 instead of question mark. Another difference using php is the use of a statement name, here test _ assay _ insert. This is used to uniquely identify the statement being prepared. The same name is used again when that statement is to be executed. Despite these differences, the principle is the same. The SQL statement is prepared once using placeholders and executed as many times as necessary to insert all data.

Placeholders are commonly used to insert or update tables using SQL. It is not possible to use a placeholder at any place in an SQL statement. For example, it is not possible to use a placeholder to represent a column or table name. It is only possible to use a placeholder in an SQL statement where a value to be inserted or updated would be used.

12.3.2   Bind Values in SQL Statements

When a client program selects data from an RDBMS table using SQL, there are several methods that can be used. The following Perl code illustrates some of these methods.

```
my $sql = "Select smiles,cas from nci.structure where gnova.
matches(smiles,'c1cccccc1C(=O)NC') limit 20";
my $sth = $dbh->prepare($sql);
my $rv = $sth->execute;
while (my @row = $sth->fetchrow_array()) {
 print join "\t",@row;
 print "\n";
}
```

In this code snippet, the SQL statement is prepared and executed and each row is fetched into an array and then printed. It is also possible to fetch all rows at one time, for example, using the following code snippet.

```
my $sth = $dbh->prepare($sql);
my $rv = $sth->execute;
my $data = $sth->fetchall_arrayref();
while ( my $row = shift(@$data) ) {
 print join "\t",@$row;
 print "\n";
}
```

After the SQL statement is executed, an array reference is returned using fetchall _ arraryref. The individual rows from this array are then printed. Some care needs to be taken when using fetchall _ arrayref when a large number of rows are returned. In that case, not all rows may be returned and the function will need to be called again until all rows are returned. The documentation for the DBI perl module discusses this more fully.[1]

In this final example, the use of bind variables is illustrated.

```
my $sth = $dbh->prepare($sql);
my $rv = $sth->execute;
$sth->bind_columns(\$smiles, \$cas);
while ($sth->fetchrow_array()) {
 print "$smiles\t$cas\n";
}
```

The bind _ columns function requires as many perl variables as there are columns in the select statement. The names used here are indicative of the columns selected, making the code more understandable. The use of bind _ columns is also very efficient.

12.4 Web Applications

Writing a Web application to help such users search or update a database is more than simply offering a text box for them to type in SQL statements. The focus of section is not to show how a full Web application can be developed that uses SQL and an RDBMS server. There are some useful SQL functions that can benefit any Web application.

Consider a Web application to allow a user to sketch in a structure, perform a substructure search, and fetch experimental data for the resulting structures. The general structure of the program is:

- Present the Web form.
- Provide for structure input (drawing).
- Provide for table search requirements.
- Process the Web form.
- Construct the appropriate SQL.
- Connect to database and select results.
- Present results to user.

Graphical input and output is essential to any chemical Web application. One common method to provide this is using Marvin or another sketching tool. Rather than have each application generate the javascript needed to present the sketcher or viewer, consider storing this code in the database and selecting it when needed. The `marvin _ sketch` function returns javascript that will cause a marvin sketch applet to appear on a Web page.

```
Create Or Replace Function marvin_sketch() Returns Text As $EOSQL$
Select '
<script LANGUAGE="JavaScript1.1">
msketch_name = "MSketch";
msketch_begin("/marvin", 200, 200);
msketch_param("background", "#EEFFDD");
msketch_param("menubar", "false");
msketch_param("molFormat", "smiles");
msketch_param("importConv", "-a");
msketch_param("detach", "hide");
msketch_end();
function get_smiles() {
 smi = MSketch.getMol("smiles:a");
 return smi;
}
</script>
'::Text;
$EOSQL$ Language SQL;
```

This SQL function can then be used by any application that needs to include a Marvin sketcher. For example, a php application would include this code.

```
<HTML>
<HEAD>
<TITLE>Sample PHP web app example</TITLE>
<SCRIPT SRC="http://www.chemaxon.com/marvin/marvin.js"></SCRIPT>
</HEAD>
<BODY>
<?php print marvin_sketch(); ?>
```

```
</BODY>
</HTML>
<?php
 function marvin_sketch() {
  $dbconn = pg_connect("host=localhost dbname=book user=reader");
  $res = pg_exec($dbconn, "select marvin_sketch()");
  $row = pg_fetch_array($res);
  pg_close($dbconn);
  return $row["marvin_sketch"];
 }
?>
```

The `marvin_sketch` SQL function can be called from any client program using database access methods described in Chapter 5. The `marvin_sketch` program can be modified to allow specification of the sketch applet size, loading an initial SMILES, or any other option provided by the Marvin sketch applet.

Most Web applications that search a database will need to know the names of the columns in a table. Rather than coding this into the application, consider writing an SQL function to provide this information. The following SQL function returns the names of the columns in a particular table.

```
Create Or Replace Function get_fields(text,text) Returns Setof
Record As $EOSQL$
Declare
 col_types Record;
Begin
 For col_types In
 SELECT attname::text, typname::text
  FROM pg_attribute
   Join pg_class On attrelid = pg_class.oid
   Join pg_namespace On relnamespace = pg_namespace.oid
   Join pg_type On atttypid = pg_type.oid
  Where attnum > 0
  And nspname = $1
  And relname = $2
 Loop
  Return Next col_types;
 End Loop;

End;
$EOSQL$ Language plpgsql;
```

This function is specific to PostgreSQL, using system catalog tables. It can be used in a Web application as in the following php example.

```
<HTML>
<HEAD>
<TITLE>Sample PHP web app example</TITLE>
<SCRIPT SRC="http://www.chemaxon.com/marvin/marvin.js"></SCRIPT>
</HEAD>
```

```
<BODY>
<FORM>
<?php
 print marvin_sketch();
 print "<p><input type=submit value=search>";
 print "pubchem.nci_h23";
 foreach (get_columns('pubchem', 'nci_h23') as $key => $value) {
   print "<BR><input type=checkbox name=${key}>${key}\n";
 }
 ?>
</FORM>
</BODY>
</HTML>
<?php
 function get_columns($schema,$table) {
  $dbconn = pg_connect("host=localhost dbname=book user=reader");
  $sql = "select * from get_fields('${schema}', '${table}') as
(colnam text, typnam text)";
  $result = pg_query($dbconn, $sql);
  $columns = array();
  while ( $row = pg_fetch_array($result) ) {
   $colnam = $row['colnam'];
   $columns[${colnam}] = $row['typnam'];
  }
  pg_close($dbconn);
  return $columns;
 }
?>
```

This expands on the previous example, adding the get _ columns php
function and the creation of the form and checkboxes showing the column names in the html body.

It is beyond the scope of this book to present an entire Web application. The purpose of this discussion is to show how creating and using
SQL functions can facilitate the creation of Web applications. The final
example shows how SMILES stored in a database can be formatted for
display using Marvin. The marvin _ view SQL function is defined as:

```
Create Or Replace Function marvin_view(text, text) Returns Text As
   $EOSQL$
Select '
<script LANGUAGE="JavaScript1.1">
msketch_name = "MView";
mview_begin("/marvin", 200, 200);
mview_param("colorScheme", "atomset");
mview_param("mol", "' || $1 || '");
mview_param("AtomSet0.1", "' ||
array_to_string(list_matches($1, $2),',') ||'");
mview_end();
</script>
'::Text;
$EOSQL$ Language SQL;
```

This function takes a SMILES and a SMiles ARbitrary Target Specifications (SMARTS). The SMARTS is used to locate a substructure within the SMILES and color the atoms that are matched.

12.5   R Programs

R is a program used to compute statistical results for sets of data. One of the commonly used data types in R is the data frame. This has many similarities to tables of data in a relational database, or tables resulting from an SQL select statement. Using the RODBC module, R can communicate with a RDMBS using SQL to read data into a data frame form and to write a data frame to an RDBMS table.

12.5.1   Hierarchical Clustering

The following example shows how R can be used to carry out a clustering analysis using data stored in an RDBMS.

```
require("RODBC");
channel = odbcConnect("PostgreSQL30", uid="reader",
case="postgresql");
sql = "Select
 Case When a.id < b.id Then tanimoto(a.gfp,b.gfp)
      Else null
 End
 from xlogp.test_set as a, xlogp.test_set as b";
tani = sqlQuery(channel, sql, max=0);
n = sqrt(length(tani[,1]))
tanimoto = as.dist(matrix(tani[,1], nrow=n, ncol=n));
fit = hclust(1.0-tanimoto, method="ward");
plot(fit);
fit = hclust(1.0-tanimoto, method="single");
plot(fit);
close(channel);
```

The SQL statement above computes the Tanimoto similarity between all pairs of compounds using fingerprint bitstrings stored in the column gfp. The tanimoto function is described in Chapter 8 and shown in the Appendix. This SQL statement uses the Case conditional clause. This is done in order to avoid computing elements unnecessarily. The matrix of similarities is symmetric and the diagonal elements are exactly 1. The sqlQuery R function reads the rows of the similarity matrix into an R data.frame named tani. This is coerced into a matrix of the correct number of rows and columns using the matrix function and further coerced into a distance R object. The R distance object is the lower half of a symmetric distance matrix. Since the tanimoto similarity is used, the distance (or dissimilarity) is represented by 1.0 minus the tanimoto value. Finally, the `hclust` R function is called using (1.0 – tanimoto). In this example, two methods are used: `ward` and `single`. Figure 12.1 shows the output from the plot function using ward clustering.

The use of the R functions is described elsewhere.[2,3] The point of this example is not to explain how to do clustering, but rather to show how easily data is read into R from an RDBMS. The full power of the SQL language and the extension functions, such as tanimoto here allows great flexibility in creating data frames for R. With very little change in the above code, various methods of clustering can be tested. With simple changes in the SQL statement above, other similarity measures, such as euclid or hamming can be investigated.

12.5.2 Linear Models

R contains a useful linear models function to carry out regression analysis to fit a set of parameters to experimental data. The example here estimates logP values using a set of fragments and coefficients, such that

$$glogp = \sum C_i * N_i \qquad \text{(Formula 12.1)}$$

where glogp is the estimated logp value for a molecule, N_i is the number of times the each fragment is contained in the molecule, and C_i is the coefficient resulting from the linear models fit. The fragments are defined as a set of SMARTS[4] to be matched against the molecule using the count _ matches function described in Chapter 7. The following R script shows how this is accomplished.

```
require("RODBC");
channel = odbcConnect("PostgreSQL30", uid="reader",
case="postgresql");

# get experimental logp from training_set
sql = "Select logp from xlogp.training_set order by id";
logpval = sqlQuery(channel, sql, max=0);
ntrain = length(logpval$logp);

# get smarts
sql = "select smarts,train_freq from xlogp.simplex
 where train_freq > 1 order by train_freq desc";
smarts = sqlQuery(channel, sql, max=0);

# match each smiles in the training_set to each fragment smarts
sql = "Select count_matches(smiles,smarts) as matches from
  (select smarts, train_freq from xlogp.simplex
   where train_freq > 1) as smarts,
  (select smiles, id from xlogp.training_set order by id) as train
   order by train_freq desc, smarts, id";
```

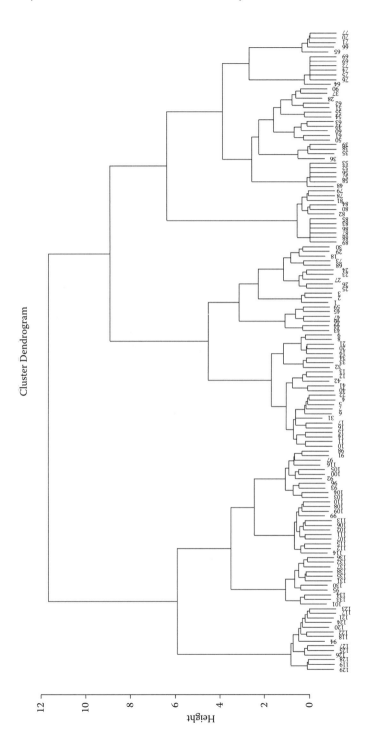

Figure 12.1 R graphical representation of Ward clustering of 1.0-tanimoto distances between compounds.

```
count = sqlQuery(channel,sql,max=0);
m = matrix(count$matches, nrow=ntrain)

# fit the experimental logP values to the matched fragment counts
logpfit = lm(data.frame(logpval, m));
summary(logpfit);
plot(fitted(logpfit), logpval[[1]], main="simplex smarts",
 ylab='experimental value', xlab='predicted value');

# create data frame of smarts and coefficients and store in a table
dt = data.frame(c(NA,as.vector(smarts$smarts)), coef(logpfit), summ
ary(logpfit)$coefficients[,2]);
names(dt) = c('smarts', 'contribution', 'error');
sqlSave(channel,dt,table='simplex_coefficients');
close(channel);
```

The first SQL statement fetches the logp values from a data table. The second SQL statement fetches a set of smarts from pre-defined atom-based fragments. The third SQL statement joins the table of smarts with a table of smiles comprising a training set and produces rows of counts of the number of times each smarts matches each smiles. The resulting rows are recast as a matrix containing rows as training_set smiles and columns as atom fragment smarts. This is combined with the logP values into a data frame passed to the lm linear models function. The results of this computation are printed by the summary function as shown below.

```
Call:
lm(formula = data.frame(logpval, m))

Residuals:
       Min        1Q    Median        3Q       Max
-2.971987 -0.375804  0.003851  0.405040  2.389759

Coefficients:
              Estimate Std. Error t value Pr(>|t|)
(Intercept) -0.262592   0.075253  -3.489 0.000496 ***
X1           0.293660   0.018785  15.633  < 2e-16 ***
X2           0.318086   0.011834  26.880  < 2e-16 ***
X3          -0.309212   0.051247  -6.034 1.94e-09 ***
X4           0.552701   0.027245  20.287  < 2e-16 ***
X5           0.324572   0.012368  26.242  < 2e-16 ***
X6           0.176798   0.045332   3.900 9.97e-05 ***
X7          -0.400717   0.033560 -11.940  < 2e-16 ***
X8          -0.313073   0.029472 -10.623  < 2e-16 ***
X9          -0.523223   0.040330 -12.974  < 2e-16 ***
X10         -0.688122   0.042941 -16.025  < 2e-16 ***
X11         -0.375329   0.034512 -10.875  < 2e-16 ***
X12         -0.058134   0.024444  -2.378 0.017497 *
X13          0.723824   0.031660  22.863  < 2e-16 ***
X14          0.008709   0.064691   0.135 0.892918
```

```
X15           0.267072    0.036639   7.289 4.63e-13 ***
X16          -0.892674    0.052076 -17.142  < 2e-16 ***
X17           0.404528    0.033359  12.126  < 2e-16 ***
X18           0.630511    0.105706   5.965 2.94e-09 ***
X19           0.316583    0.062433   5.071 4.37e-07 ***
X20           0.572682    0.160138   3.576 0.000358 ***
X21          -0.562239    0.069181  -8.127 8.05e-16 ***
X22           0.963970    0.057549  16.750  < 2e-16 ***
X23          -0.468030    0.085574  -5.469 5.15e-08 ***
X24          -0.216260    0.117513  -1.840 0.065886 .
X25          -0.241297    0.175739  -1.373 0.169909
X26           0.423008    0.079224   5.339 1.05e-07 ***
X27           0.554696    0.137816   4.025 5.93e-05 ***
X28           0.518312    0.087593   5.917 3.90e-09 ***
X29           1.375563    0.106891  12.869  < 2e-16 ***
X30          -0.224964    0.155949  -1.443 0.149320
X31           0.662818    0.187787   3.530 0.000427 ***
X32          -0.175770    0.196201  -0.896 0.370441
X33           0.825725    0.300047   2.752 0.005982 **
X34          -1.174430    0.304675  -3.855 0.000120 ***
X35           0.020394    0.371856   0.055 0.956269
---
Residual standard error: 0.6645 on 1817 degrees of freedom
Multiple R-Squared: 0.8104,     Adjusted R-squared: 0.8068
F-statistic: 221.9 on 35 and 1817 DF,  p-value: < 2.2e-16
```

This summary shows statistics on how well the counts in matrix m fit the experimental logp values. It also lists the coefficients to be used in the above equation to estimate glogp. Finally, these coefficients, along with the standard errors and SMARTS are stored in a data frame, dt. This data frame is stored in the database as a table named simplex_coefficients. The plot function produces a graph of the predicted versus experimental values as shown in Figure 12.2.

The simplex _ coefficients table can now be used in the following SQL statement to compute a glogp value for the molecule represented by the SMILES c1ccccc1C(=O)NC.

```
Select sum(contribution*count_matches('c1ccccc1C(=O)NC',smarts))
 -0.262592 as glogp from simplex_coefficients
```

The SQL aggregate function sum and the multiplication in the above statement effectively carry out the computation according to Formula 12.1 shown earlier. The constant value 0.262592 is the intercept from the lm fit shown above. A glogp function can be defined as follows.

```
Create Function glogp(text)
 Returns Numeric As $EOSQL$ Select
 (sum(contribution*gnova.count_matches($1,smarts))
 -0.262592)::numeric(5,3) From simplex_coefficients;
$EOSQL$ Language SQL;
```

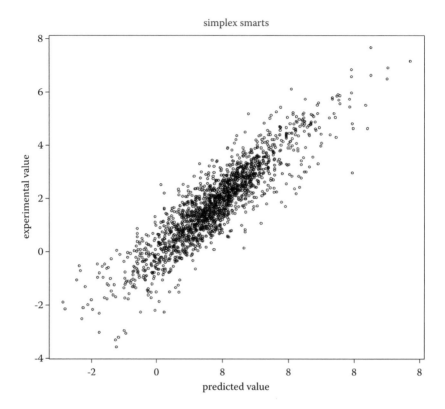

Figure 12.2 Predicted versus experimental values of logp using R linear models fit.

This function can then be used to easily estimate a logp value for any valid SMILES. For example, the following SQL computes the same result as the select statement above.

```
Select glogp('c1ccccc1C(=O)NC');
```

And the following estimates logP for an entire table.

```
Select smiles, glogp(smiles) from structure;
```

References

1. Perl   DBI   fetchall_array.   http://search.cpan.org/~timb/DBI/DBI. pm#fetchall_arrayref (accessed April 18, 2008).
2. The R Manuals. 2008. http://cran.r-project.org/manuals.html (accessed April 18, 2008).
3. Venables, W. N. and Ripley, B.D. 2002. *Modern Applied Statistics with S. 4th ed.* New York: Springer.
4. O'Donnell, T.J. 2006. Using relational databases for physical property prediction. *The 232nd ACS National Meeting*, San Francisco, CA, September 10–14, 2006.
5. Renxiao Wang, R., Fu, Y., and Lai, L. 1997. A new atom-additive method for calculating partition coefficients. *J. Chem. Inf. Comput. Sci.* 37:615–621.

chapter 13

Applications

13.1 Introduction

There are many uses for a chemical relational database. Some of these have been mentioned in earlier chapters. In this chapter, three general types of applications will be discussed. The purpose is not to present complete working applications, but to indicate important issues to consider when designing such applications. Sample schemas are proposed. The use within each application of the core functions described earlier is discussed. Each of these applications might be developed as a Web application or a client application on a user's desktop. Any computer language might be used, although the ability to connect to an RDBMS is essential.

13.2 Compound Registration

Every chemical company or research organization has a collection of compounds of interest. These may be compounds synthesized by chemists employed at the company, compounds purchased from chemical vendors, compounds on which research has been carried out, or any other collection of compounds. When a new compound becomes of interest, it is important to know whether that compound has already been entered into the system, or a new entry needs to be made. The use of canonical SMILES as a unique name for each structure makes this an easy task. One essential table in a compound registration system is a table of unique structures. Such a table could be defined as follows.

```
Drop Schema registration Cascade;
Create Schema registration;
Set search_path=registration;
  -- search_path directs following into the registration schema
Create Table structure (
 smi Text Unique Not Null,
 cansmi Text Not Null,
 id Serial Primary Key,
 fp Bit Varying);
Create Function add_new_structure() Returns Trigger As $EOSQL$
Begin
 NEW.smi    = isosmiles(NEW.smi);
 NEW.cansmi = cansmiles(NEW.smi);
```

```
 NEW.fp      = fp(NEW.smi);
 Return NEW;
End;
$EOSQL$ Language plpgsql;
Create Trigger add_new_structure Before Insert Or Update On
structure For Each Row Execute Procedure add_new_structure();
Create Index cansmi_index On structure (cansmi);
```

The first statement creates a new schema to contain these tables. The schema name is arbitrary but might be chosen to be the name of the company or research organization. Next, the `structure` table is defined to contain a `smiles` column of type `text`. This column is defined to be `unique` and `not null`. The uniqueness constraint here ensures that no duplicate compounds can be entered. The `id` column is defined using the `serial` data type. This ensures that a unique integer number will be associated with each structure. This `id` will be used in other tables within this schema to relate data in those tables to compounds in the `structure` table. The `cansmi` column will be used to contain canonical simplified molecular input line entry system (SMILES). The `fp` column will be used to contain a bit string fingerprint of the structure. The `cansmi` and `fp` column can be used when searching for compounds in this table.

The next statement defines a `trigger` function that will be used whenever data is inserted or updated in this table. This function performs three important functions. First, it modifies the SMILES to be inserted into the `smi` column so that it contains the result of the `isosmiles` function. The `isosmiles` function is similar to the `cansmiles` function, except that it retains any stereochemistry that might be contained within the SMILES. If two stereoisomers are entered into this table, each will have a unique `isosmiles` value, but the same `cansmiles` value. In this way, they can be kept distinct, but their identical canonical SMILES shows them to be stereoisomers. The `trigger` function also computes the fingerprint and inserts it into the table when the SMILES is inserted or updated.

The use of the uniqueness constraint and the `trigger` ensures integrity of the data in this table. However, it does not automatically correct all problems. For example, if the insertion of invalid SMILES is attempted, an error will be generated and the SMILES will not be inserted. It is the responsibility of the application program to deal with this error. If the application is a batch style application, there error must certainly be logged so that it can be dealt with at a later time. If it is an interactive application, the user is informed of the error and asked to re-enter the structure. If the SMILES is valid, the `isosmiles`, `cansmiles`, and `fp` function should not fail. However, if there is some fault in those functions, the overall insert or update will not occur. For this reason, it might be desirable to log any error that occurs. These errors might be available in the log file of the RDBMS server, depending on how that server is configured. Alternatively, a table of failed SMILES can be maintained. This can be accomplished by modifying the add _ new _ structure function as follows.

```
Create Table error_log (smi text,
 attempt timestamp(0) Default current_timestamp );
Create Function add_new_structure() Returns Trigger As $EOSQL$
Begin
 NEW.smi     = isosmiles(NEW.smi);
 NEW.cansmi = cansmiles(NEW.smi);
 NEW.fp      = fp(NEW.smi);
 Return NEW;
Exception
 When OTHERS Then
  Insert Into error_log (smi) Values (NEW.smi);
  Return Null;
End;
$EOSQL$ Language plpgsql;
```

The use of the exception clause traps errors in any of the isosmiles, cansmiles, or fp functions. The creation of the error _ log table is shown above to contain these errors for later inspection and correction.

The id column is defined as a primary key. This causes an index to be created, which will facilitate joining the structure table with other tables yet to be created. The smiles column is defined to be unique, which also automatically creates an index. This column will not be used as a key, but the unique index will allow fast lookups on this table if a particular structure is desired. The final definition of this schema creates an index on the cansmiles column. This will not be a unique index, but it will allow fast lookup of structures by canonical SMILES.

Searching for structures in the structure table can be done in many ways, but several important methods are discussed here. First, a structure can be located directly using the following structured query language (SQL).

```
Select id, smi from structure Where cansmi=cansmiles
  ('c1ccccc1C(=O)NC'));
```

The actual value of the SMILES, here c1ccccc1C(=O)NC would come from the user, perhaps using a drawing widget or other method. The use of the cansmiles function assures that the SMILES string value will correspond to a value stored in the cansmiles column. Recall that the cansmi column was populated using the cansmiles function in the trigger function shown above. If the SMILES input by the user is an isomeric SMILES, the SQL above will locate all isomers. Related stereoisomers all share the same canonical SMILES. If it is desired to locate only the isomeric SMILES, the following SQL could be used.

```
Select id, smi From structure Where smi=isosmiles('F[C@H](C)Cl'));
```

If a substructure search is desired, it is wise to use the fingerprint stored in the fp column to reduce the number of structures that must be scanned using the matches function. The following SQL will locate all structures that contain the specified substructure.

```
Select id,smi From structure Where
 contains(fp, fp('clccccc1C(=O)NC')) And
    matches(smi, 'clccccc1C(=O)NC'));
```

The addition of the contains function allows a quicker comparison of the fingerprint of the desired substructure with the fingerprints stored in the table. The matches function is then used only for structures which have passed this initial test. Since the matches function is slower than the contains function, the overall speed of the search is faster than if the fingerprint comparison were not done.

It might be tempting to add additional columns to the structure table to hold defined properties of each structure. Not all properties of a structure are appropriate for a table of structures. Some properties, for example, molecular weight and molecular formula are fixed properties of a structure with a unique value. These might be added as columns to the structure table. However, they could also be kept in another table related to the structure table. Consider also how often these values will be needed or if they will be searched. It is possible to easily compute these properties when needed, using SQL functions that take a SMILES argument.

Other properties are not unique, for example, chemical names. These should be stored in a separate table with one row for each value. For example, the entry in the pubchem database contains 10 synonyms for the SMILES C1(C(C(C(C(C1O)O)OP(=O)(O)O)O)O)O as shown in Table 13.1. Each of these should be entered as a separate row in a table of names along with a column containing the compound id. A simple table of this type would be created using the following SQL.

```
Create Table names (cid integer References structure (id), name text);
```

The cid column is a foreign key referencing the id column of the structure table. This prevents any names from being entered that do not have a corresponding entry in the structure table. It also associates the name with the proper structure. As shown in earlier chapters, names, and smiles can be selected from the tables in this schema using the following SQL.

```
Select smi, name From structure Join names On (id=cid);
```

Any number of other tables can be added to this schema. Each should be related to the structure table using the compound id. Aside from simply registering compounds, it might be required to store experimental data

Table 13.1 Sample of 10 Synonyms
in pubchem for inosotol 1-phosphate

Name
1-L-MYO-INOSITOL-1-P
1D-myo-inositol 3-monophosphate
1D-myo-inositol 3-phosphate
1L-myo-inositol 1-phosphate
D-myo-inositol 3-phosphate
L-myo-inositol 1-phosphate
inositol 1-phosphate
myo-inositol 1-monophosphate
myo-inositol 1-phosphate

about structures in the database. Most properties, such as experimental values may have multiple values depending on how they were measured. Experimental values belong in a separate table where information about those values can be stored as well as the values themselves. For example, the date of the measurement, uncertainties, methods, as well as multiple values might be stored. When storing multiple values for any one structure, each value is stored as a separate row in the table with the same structure id. The reasons for doing this were discussed in Chapter 2 in the section on normal forms. It is possible to create additional tables in the same schema, but this is not necessary. It might be useful to create another schema for each type of experimental data, for example for each assay or project. The way to store experimental data is discussed in a later section of this chapter.

This approach for creating a registry of compounds might be expanded in many ways. One important thing to consider is how tautomers are handled. For some tautomers, a set of rules can be devised. In principle, all tautomeric forms of any compound are valid and detectable chemical entities. But for some compounds, one tautomeric form is so dominant, or the other forms are of so little interest, that one form can be chosen as the standard tautomer. For example, all nitro groups could be stored as charge-separated, for example C[N+](=O)[O−] or as CN(=O)=O. Unless your projects are concerned with the very topic of tautomerization in nitro groups, this is usually an arbitrary choice. It is usually helpful to enforce a standardization rule to create a unique set of structures that can be accurately searched. Rather than forcing users to remember the rules, a table of transformations is used to change structures entered with the "wrong" form into the standard form. This could be done using the transformation and reaction functions discussed in Chapter 9. In order to accomplish this, the add _ new _ structure trigger function shown above could be expanded as follows.

```
Create Function add_new_structure() Returns Trigger As $EOSQL$
Declare
  std_smiles Text;
  smirks Text;
  std Record;
Begin
  For std In Select * from std_smirks Loop
    std_smiles = xform(NEW.smi, std.smirks);
    If std_smiles != NEW.smi Then
      NEW.smi = std_smiles;
    End If;
  End Loop;
  NEW.smi     = isosmiles(NEW.smi);
  NEW.cansmi  = cansmiles(NEW.smi);
  NEW.fp      = fp(NEW.smi);
  Return NEW;
Exception
  When OTHERS Then
    Insert Into error_log (smi) Values (NEW.smi);
    Return Null;
End;
$EOSQL$ Language plpgsql;
```

The std_smirks table would contain the standard transformations. Chapter 9 shows a sample table of standard SMIRKS transformations. This table could be expanded at any time to include more standardizations as they become necessary without having to modify the trigger function to deal with these additions.

Other types of tautomers are not so easy to standardize. Some tautomers are not arbitrarily different ways of "spelling" a SMILES. These tautomers are readily isolated chemical entities for which different chemical properties can be measured. For example, dihydroxynapthelene exists in two forms[1] as shown in Figure 13.1. When the dihydroxy form (Oc1ccc(O)c2ccccc12) is registered, it must be assumed that the dihydroxy form is intended rather than the diketo form (O=C3CCC(=O)c4ccccc34) of the compound. However, it is possible to detect that a tautomer of a compound is already present in the database. Using the graph function described in

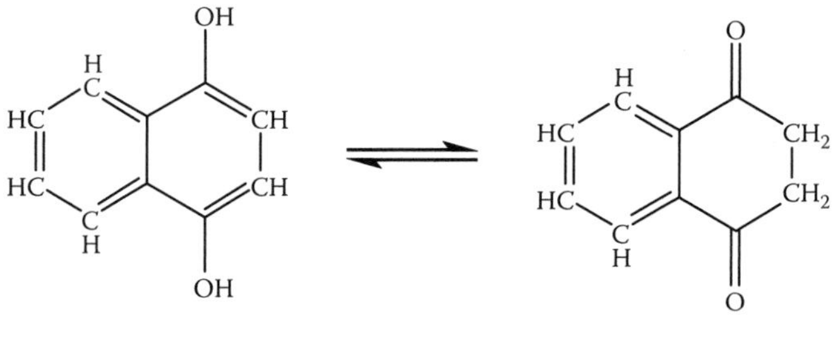

Figure 13.1 Two enantiomeric forms of dihydroxynapthelene.

Chapter 7, it is clear that these two structures have the same simple graph and are therefore tautomers. The following SQL returns true.

```
Select graph('Oc1ccc(O)c2ccccc12') = graph('O=C3CCC(=O)c4ccccc34');
```

In order to facilitate tautomer detection, a column of graphs of each structure could be stored and searched whenever a new structure is entered. The user entering the structure could be asked if the tautomeric form already in the database is perhaps the intended form for this structure. Even if the new tautomeric form is entered as a new compound, the tautomer relationship between the two structures will be recorded by virtue of them having the same graph. The process of creating and comparing graphs can be added to the add _ new _ structure trigger shown above. The modified function and the modified table and associated index is shown below.

```
Create Table structure (
 smi Text Unique Not Null,
 cansmi Text Not Null,
 grf Text Not Null,
 id Serial Primary Key,
 fp Bit Varying);
Create Index grf_index On structure (grf);
Create Function add_new_structure() Returns Trigger As $EOSQL$
Declare
 std_smiles Text;
 smirks Text;
 std Record;
Begin
 For std In Select * from std_smirks Loop
  std_smiles = xform(NEW.smi, std.smirks);
  If std_smiles != NEW.smi Then
    NEW.smi = std_smiles;
  End If;
 End Loop;
 NEW.smi    = isosmiles(NEW.smi);
 NEW.cansmi = cansmiles(NEW.smi);
 NEW.grf    = graph(NEW.smi);
 NEW.fp     = fp(NEW.smi);
 Return NEW;
Exception
 When OTHERS Then
  Insert Into error_log (smi) Values (NEW.smi);
  Return Null;
End;
$EOSQL$ Language plpgsql;
```

This addition does not correct the issues with tautomers, but it does allow an easy way to detect tautomers in the database. Note also that alerting the user is the responsibility of the client program and is not performed in this trigger function or in any of the other constraints in the registration schema.

There is some overhead in the use of indexes, constraints, triggers, etc. as discussed here. The overhead is incurred when rows are inserted or updated in the table. However, the value of this approach is that the data in the table are well validated and can be searched more reliably and efficiently. Direct lookups of canonical or stereo SMILES is simple and quick because of the index on these columns. Using the fingerprint column speeds up substructure search. Tautomers can be readily selected using the column of simple graphs.

It might be helpful to delay the creation of the indexes when the schema is first created and its tables populated. This is especially true if millions of compounds are to be entered at one time. However, if there are duplicate structures and the table contains even two rows with the same isosmi, it will not be possible to create a unique index on the isosmi column until only a unique set of isosmi values exists. The creation of a unique index does not fix nonunique values. It simply prevents non-unique values. In order to find duplicate structures in a table, the following SQL can be used.

```
Select isosmiles, count(isosmiles) from structure
 group by isosmiles having count(isosmiles) > 1;
```

The use of the group by clause causes all identical values of isosmiles to be grouped together and processed by the aggregate function count. For duplicate values of isosmiles, the count will be greater than 1. The above SQL statement will select these isosmiles.

13.3 Experimental Chemical and Biological Data Integration

The first section of this chapter showed how a schema of tables could be used to create a compound registry. Using that schema, this section will show how experimental data can be integrated with compound data. A separate schema will be used to store the experimental data. In fact, several schemas will be created in order to segregate data tables for separate assays or projects. This is not essential, but is handy for browsing data tables in a large database. Schemas are analogous to folders in a file system.

Each set of experimental data will likely have to be considered separately. Some measurements may require several columns, while other will require only one. Some measurements may have measured or estimated uncertainties, while others may be more indefinite, such as "active" or "inactive." All measurements will have a date associated with them and this should be stored. Many experimental values will be noninteger numbers. These may be stored using the numeric data type or float or double, depending on the accuracy needed and whether computations on these values will be carried out from within the database. For example, the `numeric` data type does not truncate or round the input values and keeps as many significant figures as there are in the external representation of the number when it is entered. Using the `float` or `double` internal representation will result in slight rounding of some input values. Computations carried out with `float` values are more efficient that computations using `numeric` values. So, if the values in the table will be used with many mathematical functions, such as `sqrt`, `average`, `standard deviations`, etc. and if extreme accuracy is not crucial, using a `float` data type may be preferred.

A table of experimental data could be defined as follows.

```
Create Schema hiv;
Create Table hiv.protease (id Integer References registry.structure,
ic50 Float, ec50 Float, updated Timestamp Default current_timestamp);
```

This table is intended to hold results of assays testing compounds in registry.structure for activity as human immunodeficiency virus (HIV) protease inhibitors. As new assays are added, the test results can be added to newly created tables with similar definitions. For example, there might be tables for HIV reverse transcriptase inhibitors stored in a table named `hiv.rt`. Other assay results might be stored in new schemas, for example, `fpr.htfc` for high-throughput flow cytometry results for the formyl peptide receptor (FPR), or `fpr.ca` for FPR cell adhesion assay results. Each of these tables would have columns of data named and typed appropriately for each assay. Each table would have a column containing a compound id that references compounds in the `registry.structure` table.

It is not possible to propose a schema with tables that can accommodate experimental results of any type. It is important to consider the needs of each project and assay so that appropriate tables can be created with the necessary data types and constraints. One common feature of any table of experimental data is a column containing a reference to a chemical compound or compounds involved in the experimental measurement. While the examples so far have considered only one compound for each test result row, it is important to consider how results will be handled when multiple compounds are involved in each experimental measurement, or when multiple measurements are made for the same compound with samples prepared at different time or perhaps in different ways. A common way to handle these situations is to use the concept of a sample.

A sample can be defined as a preparation consisting of one or more compounds and one or more solvents. It might also be necessary to consider separate batches of any compound, obtained at different times from different vendors or synthesized at different times. Using samples and sample ids instead of compounds and compound ids is a more accurate reflection of the actual experimental situation and can record more information about how the measurements were made. In order to accommodate the use of samples, it is necessary to create a sample table and use sample ids in the experimental data tables in place of compound ids. A sample table could be created as follows.

```
Create Sequence uniq_samples;
Create Table sample (id Integer Default Nextval('uniq_samples'),
 cid Integer References registry.structure (id),
 prepared Timestamp(0) Default current_timestamp,
 Unique (id, cid));
```

This introduces a new way in which sequences are used within a database. In previous chapters, the `Serial` data type was used to create a column of unique integers. When the `Serial` data type is used, a sequence is automatically created and the default value of the column is set to be the next value in the sequence. In this way, a unique set of integers is ensured. In the above example, more control is needed over the use of sequence values in the `sample.id` column. This is because this table may contain several rows with the same sample `id`, each with a different compound `id`. The sample consists then of all compounds in the sample table having the same sample `id`.

When a new sample is inserted into the `sample` table, only the `sample.cid` need be specified. The default value for `id` will cause a new `sample.id` to be taken from the `uniq _ samples` sequence. The default value for `sample.prepared` will cause the current time to be inserted into that column. If a further specification of the same sample needs to be made, the following SQL will suffice to add compound id 55.

```
Insert into sample (id, cid) Values (currval('uniq_samples'), 55);
```

The currval function uses the current value of the sequence. This can be repeated as many times as necessary to complete the definition of the sample with that `sample.id`.

The final clause in the create statement above demands that the combination of `sample.id` and `sample.cid` be unique. This allows many rows with the same `sample.id`. Each sample can contain many compounds. It also allows many rows with the same `sample.cid`. Each compound can be a part of many samples. It forbids the same `sample.id` from containing the same compound (`sample.cid`) more than once.

13.4   Data from External Sources

When creating a schema of tables for projects underway at your company or research institution, there is complete freedom to define the tables as necessary to accommodate the data correctly. Sometimes it is necessary or desirable to import data from an external source. In this case, a careful understanding of the data values and relationships among them is necessary. Often the data will be made available in flat files with little or no structure. It is important to review these files and create a schema of tables appropriate for these data. Chapter 6 showed how this might be done for data obtained from the PubChem project.[2] The concepts of unique compound ids and sample ids was used there, along with a separate schema for assay results containing references to sample ids. Chapter 11 showed an example of importing data on VLA-4[3] Integrin antagonists with a simpler data organization. In both cases, these were imported into new schema designed to fit those data. It may be desirable to integrate newly imported data into existing schemas in the database. For example, if the new data contains information about compounds already in the registry tables, that relationship should be recorded. Similarly, if the imported data contains compounds not already in the registry, they could be added to the registry.

The following SQL function returns a compound id that can be used when processing data to be imported from an external source. The function returns the structure.id of an existing compound, or creates a new entry in the structure table for a new compound if necessary and returns that new `structure.id`.

```
Create Or Replace Function registry.compound_id (Text) Returns
Integer As $EOSQL $
Declare
 cid Integer;
Begin
 Select id Into cid from structure Where smi = isosmiles($1);
 If cid Is Null Then
  Insert into structure (smi) Values ($1);
  Select id Into cid From structure Where smi = isosmiles($1);
 End If;
 Return cid;
End;
$EOSQL$ Language plpgsql;
```

This function is placed into the `registry` schema and would be called as in the following example.

```
Select registry.compound_id('SCCS');
```

As compounds are processed from the external source, say from an sdf file, each value of the SMILES would be passed to the `compound_id` function and the resulting id would be used in the column of the new table that references the `registry.structure` id column.

As compounds are added to the registry from various sources, it may be necessary to record where these compounds came from. This introduces another generally useful feature in a registry of compounds. An arbitrary collection of structure ids is often useful. As compounds are added from an external source, they form such a collection. A chemist may create other collections for any purpose. The following tables record lists of registered compounds.

```
Create Table list_description (id Serial, details Text);
Create Table lists (id Integer References list_description (id),
  cid Integer References structure (id));
```

The list _ description table is updated when a list is created. In the case of importing compounds from an external source, the list _ description.details column might contain the file name, URL, and any other information about the source of the compounds. As each compound is processed using the compound _ id function, the compound id is added to the lists table using the list _ description.id and the compound id. It is possible for a compound to be contained in the registry.lists table many times. This reflects the fact that the compound is used in many lists, available from many sources, imported from many external sources or of interest in many projects. Figure 13.2 shows an entity relationship diagram incorporating the tables discussed here. There will be many other tables of experimental data in other schemas that are related to the registry.structure table.

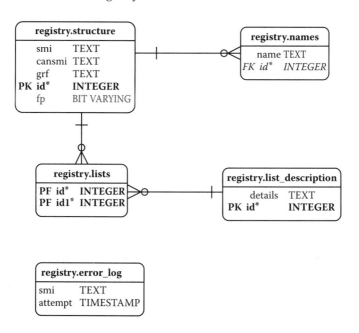

Figure 13.2 Entity relationship diagram for schema containing tables for a compound registration system.

13.5 Utilities

In any project, there will be collections or files of compounds that need to be processed. This chapter and previous chapters have shown ways in which these can be usefully imported into a database. Traditionally these files are processed in some way without being imported into a database. There are many utility functions to carry out operations such as locating structures within a file, finding nearest neighbors, clustering compounds, displaying common substructures, etc. These often take the form of command line tools, or methods within a programming environment such as python. If these tools are collected together and placed as functions in an RDBMS, these utilities can be used from within the database. They can also be used as command line tools, or integrated into a programming environment. This section will show how some of these operations can be carried out as command line utilities.

Every one of these utilities will first require that a file of structures be loaded into a table in the database. Two methods are shown here: importing a SMILES file and a mol file. Other file types could be added as needed, extending the core functions described earlier using `molfile_mol` or `molfile_to_smiles` as a model. OpenBabel is a good choice because of its support of many file formats.

A SMILES files is readily imported into the database using the following perl script `smiloader`. The output of this command is a set of SQL commands interspersed with lines in the input SMILES file. The file is minimally processed. The script expects the name of a schema in which the tables will be created. The entire perl script is shown in the Appendix. It is used at the linux command line as follows. The schema name here is `drugs`, the first argument to `smiloader`.

```
perl smiloader drugs <drugs.smi | psql mydb
```

The SQL output is piped to the psql command that process the commands. The schema and tables are created in a database named `mydb` in this example. If no database name is given, psql assumes a database with the same login name as the user. The table created by `smiloader` contains four columns: name `text`, id `integer`, isosmiles `text`, and fp `bit varying`.

Loading a SDF file is similar, although additional tables are created to accommodate the data items in the file and to contain the original file. The `sdfloader` script is used as in the following example.

```
perl sdfloader vla4 <vla-4.sdf | psql mydb
```

Figure 13.3 shows an entity relationship diagram for the tables created by the `sdfloader` script. Once files have been loaded into the database, the `dbutils` shell script is run to define several utility functions that operate on these tables. The `dbutils` script is listed in the Appendix. This

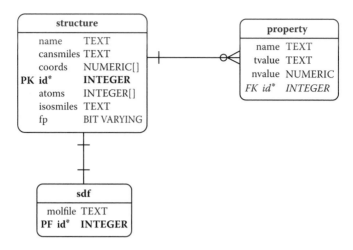

Figure 13.3 Entity relationship diagram for table created by sdfloader.

script defines the following commands: `molgrep`, `molcat`, `molview`, `molarb`, `molrandom`, `molnear`, and `molsame`. These are described individually below.

13.5.1 *molgrep*

The `molgrep` utility takes two arguments: the name of the schema containing structures previously loaded, and a SMILES or SMARTS string. The result is a list of SMILES and names from `structure` table, which match the SMARTS string or which contain the SMILES as a substructure. For example:

```
molgrep drugs 'c1cccccc1C(=O)NC'
```

prints the following subset from the `drugs.structure` table.

```
OC(=O)c1ccccc1Nc1cccc(c1)C(F)(F)F|flufenamic
Fc1ccc(cc1)C(=O)CCCN1CCC(O)(CC1)c1ccc(Cl)cc1|haloperidol
CCN(CC)CCNC(=O)c1ccc(N)cc1|procainamide
CCCCNc1ccc(cc1)C(=O)OCCN(C)C|tetracaine
```

13.5.2 *molcat*

The `molcat` utility takes one argument: the name of the schema into which structures were loaded using the `smiloader` or `sdfloader` utility. It prints the SMILES and names of all the structures in the `structure` table in that schema. For example:

```
molcat vla4
```

13.5.3 *molview*

The `molview` utility takes two arguments: the name of the schema containing structures previously loaded, and a SMILES or SMARTS string. Like the `molgrep` command, the result is a list of SMILES and names that match the SMARTS string or that contain the SMILES as a substructure. However, the results are formatted using HTML and javascipt to call the Marvin applet to display the structures. The atoms in each structure that match the SMARTS or SMILES argument are colored. If no SMILES or SMARTS argument is given, the structures in the entire table are output. The command

```
molview vla4 'c1ccccc1C(=O)' >mtest.html
```

results in the following display in a Web browser. Note that the Marvin applet[1] must be downloaded and made available to the Web server being used. Figure 13.4 shows the output as viewed in a Web browser.

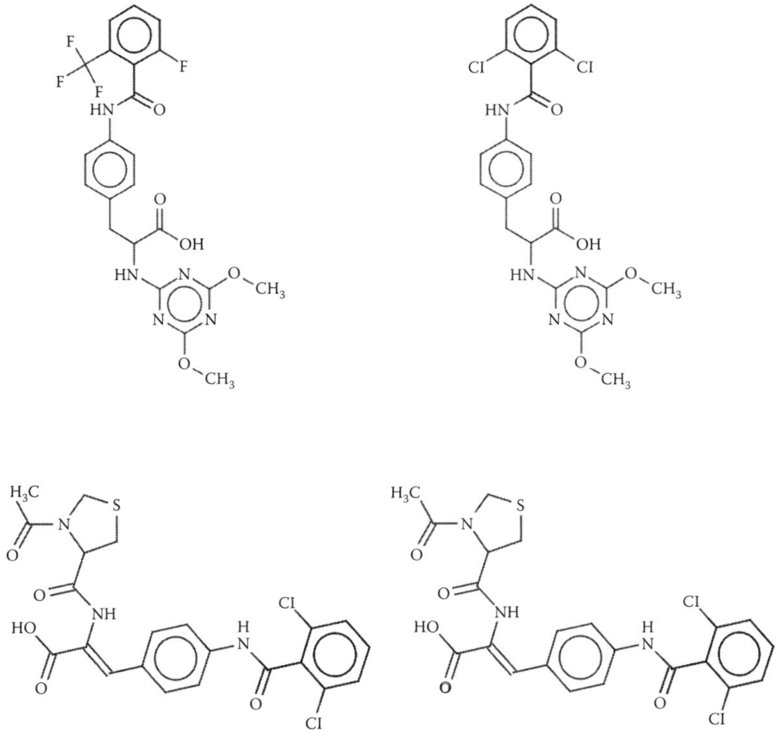

Figure 13.4 Structures displayed using Marvin and output from molview utility command.

13.5.4   *molarb*

The `molarb` utility returns an arbitrary set of structures from the structure table in a schema. There are many sets that can be devised. Each one is identified by a set number. Each set consists of one half of the table of structures. The structures in the table are segregated into two arbitrary sets based on the structure `id` number by using the `md5` function to hash the `id` number with the set number. The `molarb` utility takes three arguments: the name of the schema containing structures previously loaded, and two numbers. The first number is the number of structures desired in the output. The second number is the set number desired. Each time the `molarb` utility is used on the same table, the same set number produces the same set of structures. If a random set of structures is desired, use the `molrandom` utility. Sample output from the `molarb` utility is shown below.

```
> molarb vla4 5 2
OC(=O)C(Cc1ccc(cc1)NC(=O)c1c(Cl)cncc1Cl)NC1=C(N2CCOCC2)C(=O)C1=O|BMCL-1051-36
OC(=O)C(Cc1ccc(cc1)NC(=O)c1c(Cl)cncc1Cl)Nc1ncnc(Cc2ccccc2)c1|BMCL-1595-09
CCCSc1ncnc(c1)NC(Cc1ccc(cc1)NC(=O)c1c(Cl)cncc1Cl)C(=O)O|BMCL-1595-06
Clc1ncnc(c1)NC(Cc1ccc(cc1)NC(=O)c1c(Cl)cncc1Cl)C(=O)O|BMCL-1595-01
COc1cc(NC(Cc2ccc(cc2)NC(=O)c2c(Cl)cncc2Cl)C(=O)O)nc(n1)S(=O)(=O)
  C|BMCL-1595-15
> molarb vla4 5 8
OC(=O)C(Cc1ccc(cc1)NC(=O)c1c(Cl)cncc1Cl)NC1=C(NCc2ccccc2)C(=O)
  C1=O|BMCL-1051-30
COc1nc(OC)nc(n1)NC(Cc1ccc(cc1)NC(=O)c1c(Cl)cncc1Cl)C(=O)O|BMCL-1591-4A
COc1nc(OC)nc(n1)NC(Cc1ccc(cc1)OCc1c(Cl)cccc1Cl)C(=O)O|BMCL-1591-25
COc1nc(OC)nc(n1)NC(Cc1ccc(cc1)NC(=O)c1c(Cl)cccc1Cl)C(=O)O|BMCL-1591-24
OC(=O)C(Cc1ccc(cc1)NC(=O)c1c(Cl)cncc1Cl)Nc1ncnc(Cc2ccccc2)c1|BMCL-1595-09
```

13.5.5   *molrandom*

The `molrandom` utility returns a random set of structures from the `structure` table in a schema. The set consists of one half of the table of structures. The structures in the table are segregated into two arbitrary sets based on structure `id` and by using the `md5` function to hash the structure `id` with a random number. This has the same result as using a random number for the set number used in `molarb`, described above. The `molrandom` utility takes two arguments: the name of the schema containing structures previously loaded, and the number of structures desired. Sample output from molrandom is shown below.

```
> molrandom vla4 5
OC(=O)C(Cc1ccc(cc1)NC(=O)c1c(Cl)cncc1Cl)NC1=C(NCCCC(F)(F)F)C(=O)
  C1=O|BMCL-1051-20
OC(=O)C(Cc1ccc(cc1)NC(=O)c1c(Cl)cncc1Cl)NC1=C(N2CCSCC2)C(=O)
  C1=O|BMCL-1051-37
CCCS(=O)(=O)c1ccnc(NC(Cc2ccc(cc2)NC(=O)c2c(Cl)cncc2Cl)C(=O)O)
  c1|BMCL-1595-29
```

```
COc1nc(nc(n1)N(CC)CC)NC(Cc1ccc(cc1)NC(=O)c1c(Cl)cncc1Cl)C(=O)
   O|BMCL-1591-17
C=CCNC1=C(NC(Cc2ccc(cc2)NC(=O)c2c(Cl)cncc2Cl)C(=O)O)C(=O)
   C1=O|BMCL-1051-18
```

13.5.6   molnear

The `molnear` utility outputs a set of structures that are similar to a given structure. It takes three arguments: the name of the schema containing a table of structures, a SMILES string representing the reference structure, and a similarity index number between 0.0 and 1.0. The fingerprint in the structure table and the fingerprint of the reference structure are compared using the tanimoto function. The tanimoto function is described in the appendix. Sample output from the `molnear` command is shown below.

```
> molnear vla4 'c1ccccc1C(=O)NC' 0.2
COc1nc(nc(n1)N(C)C)NC(Cc1ccc(cc1)NC(=O)c1c(Cl)cncc1Cl)C(=O)
   O|BMCL-1591-16|0.201439
COc1nc(nc(n1)N(CC)CC)NC(Cc1ccc(cc1)NC(=O)c1c(Cl)cncc1Cl)C(=O)
   O|BMCL-1591-17|0.2
CCCNc1nc(OC)nc(n1)NC(Cc1ccc(cc1)NC(=O)c1c(Cl)cncc1Cl)C(=O)
   O|BMCL-1591-18|0.201439
OCCNc1nc(OC)nc(n1)NC(Cc1ccc(cc1)NC(=O)c1c(Cl)cncc1Cl)C(=O)
   O|BMCL-1591-19|0.201439
COc1nc(OC)nc(n1)NC(Cc1ccc(cc1)NC(=O)c1c(F)cccc1C(F)(F)F)C(=O)
   O|BMCL-1591-23|0.223022
COc1nc(OC)nc(n1)NC(Cc1ccc(cc1)NC(=O)c1c(Cl)cccc1Cl)C(=O)
   O|BMCL-1591-24|0.244094
COc1nc(OC)nc(n1)NC(Cc1ccc(cc1)N(C)C(=O)c1c(Cl)cncc1Cl)C(=O)
O|BMCL-1591-26|0.202899
OC(=O)C(NCC(=O)c1c(Cl)cncc1Cl)Cc1ccc(cc1)NC(=O)c1c(Cl)
cncc1Cl|BMCL-1591-3|0.237288
COc1nc(OC)nc(n1)NC(Cc1ccc(cc1)NC(=O)c1c(Cl)cncc1Cl)C(=O)
O|BMCL-1591-4|0.202899
Clc1ncnc(c1)NC(Cc1ccc(cc1)NC(=O)c1c(Cl)cncc1Cl)C(=O)
O|BMCL-1591-9|0.201439
Clc1ncnc(c1)NC(Cc1ccc(cc1)NC(=O)c1c(Cl)cncc1Cl)C(=O)
O|BMCL-1595-01|0.201439
Clc1ccc(nc1)NC(Cc1ccc(cc1)NC(=O)c1c(Cl)cncc1Cl)C(=O)
O|BMCL-1595-25|0.215385
```

13.5.7   molsame

The `molsame` utility outputs structures that are in common between two `structure` tables. It takes two arguments: the names of the two schemas containing a structure table. Sample output is shown below.

```
> molsame drugs pubtest
CN(C)CCCN1c2ccccc2Sc2ccc(Cl)cc12|chlorpromazine
Clc1ccc2N(C)C(=O)CN=C(c3ccccc3)c2c1|diazepam
```

```
Fc1ccc(cc1)C(=O)CCCN1CCC(O)(CC1)c1ccc(Cl)cc1|haloperidol
CN(C)CCCN1c2ccccc2CCc2ccccc12|imipramine
CCC1(C(=O)NC(=O)NC1=O)c1ccccc1|phenobarbital
CCN(CC)CCNC(=O)c1ccc(N)cc1|procainamide
OC(CNC(C)C)COc1cccc2ccccc12|propranolol
CCCCNc1ccc(cc1)C(=O)OCCN(C)C|tetracaine
COc1cc(Cc2cnc(N)nc2N)cc(OC)c1OC|trimethoprim
Coc1ccc(CCN(C)CCCC(C#N)(C(C)C)c2ccc(OC)c(OC)c2)cc1OC|verapamil
```

The `pubtest` schema was created using the first 25,000 substances from pubchem, namely Substance_00000001_00025000.sdf available from the PubChem project.[5]

References

1. Kündig, E.P., García, A.E., Lomberget, T., and Bernardinelli, G. 2006. Rediscovery, isolation, and asymmetric reduction of 1,2,3,4-tetrahydronaphthalene-1,4-dione and studies of its [Cr(CO)3] complex. *Angew. Chem. Internat. Ed.* 45(1):98–101.
2. Pubchem at http://ncbi.nih.gov/pubchem/ (accessed April 18, 2008).
3. Porter, J.R., Archibald, S.C., and Brown, J.A. 2003. Dehydrophenylalanine derivatives as VLA-4 integrin antagonists. *Bioorg. Med. Chem. Lett.* 13(5):805–808.
4. Marvin. http://www.chemaxon.com/ (accessed April 18, 2008).
5. FTP directory /pubchem at ftp.ncbi.nih.gov. 2008: pubchem. ftp://ftp.ncbi.nih.gov/pubchem/ (accessed April 18, 2008).

Appendix

A.1   Introduction

This Appendix contains structured query language (SQL) functions and tables too large or complex for the explanatory nature of the earlier chapters. These functions and tables are practical, rather than explanatory. They all follow PostgreSQL syntax. Some of them require the core functions described in Chapter 7 of this book, for example, match, cansmiles, and count _ matches. Those functions are available in the CHORD product from gNova, Inc. This Appendix also contains a PerlMol implementation, a FROWNS implementation, and an OpenBabel implementation of the core functions for PostgreSQL.

A.2   Symbols and Bonds from Simplified Molecular Input Line Entry System (SMILES)

If SMILES is used to store molecular structures in a relational database management system(RDBMS), it may be necessary to extract the symbol and bond information for some client programs that expect a connection table. The smiles _ to _ symbol and smiles _ to _ bonds function shown in the next sections allow the symbol and bond information in a SMILES to be extracted as an array. Some client programs may prefer to process this information in rows, as if they were records in a file. The following plpgsql functions can be used to present the array elements as rows. Two functions are shown: ctable (connection table) and symbol_ coords. The symbol _ coords function requires an array of coordinates in addition to the symbols.

```
Create or Replace Function ctable(text) Returns Setof Record As
  $EOSQL$
-- Called with $1 as SQL selecting integer[] as bonds
-- Example caller:
-- select * from
--   ctable('select smiles_to_bonds(cansmiles) as bonds
--   from vla4.structure where name=''BMCL-1051-38''')
--   as (atom1 integer, atom2 integer, bond_order integer);

  Declare
   bonds Record;
   b      Record;
   i      Integer;
  Begin
   For b In Execute $1 Loop
    For i in 1 .. array_upper(b.bonds,1) Loop
     Select b.bonds[i][1], b.bonds[i][2], b.bonds[i][3] Into bonds;
     Return Next bonds;
    End Loop;
   End Loop;
  End;
$EOSQL$ Language plpgsql;

Create or Replace Function symbol_coords(text) Returns Setof Record
  As $EOSQL$
-- Called with $1 as SQL selecting text[] as symbols, numeric[] as
  coords
-- Example caller:
-- select * from
--   symbol_coords('select smiles_to_symbols(cansmiles) as symbols,
--   coords from vla4.structure where name=''BMCL-1051-38''')
--   as (symbol text, x numeric, y numeric, z numeric);

  Declare
   sym_coord Record;
   sc        Record;
   i         Integer;
  Begin
   For sc In Execute $1 Loop
    For i in 1 .. array_upper(sc.symbols,1) Loop
     Select sc.symbols[i], sc.coords[i][1], sc.coords[i][2],
       sc.coords[i][3]
         Into sym_coord;
     Return Next sym_coord;
    End Loop;
   End Loop;
  End;
$EOSQL$ Language plpgsql;
```

These functions are called with a string argument. This argument is an SQL statement that is expected to provide the required information. For ctable, this is an array of bonds. For symbol _ coords, these are an array of symbols and an array of coordinates for each atom.

A.3 Normalizing Data

When a data value is repeated multiple times in a column in a database, it is said to violate third normal form. For example, a table of values for logp might contain a column named ref having literature references. The value 'Hansch, et. al. (1995)' might be repeated many times. It is easy to spot this, and easy to correct it as well. The following SQL can be used to help put a table of logp values and references into third normal form.

```
-- New table to hold all references identified by unique id
Create Table literature_refs (refid Serial, reference Text);
-- Populate table with unique references found in logp table
Insert Into literature_refs (reference)
  Select Distinct ref From logp Group By ref;
-- Create column in logp to hold reference id instead of reference
Alter Table logp Add Column refid integer;
-- Populate logp table's reference id column with appropriate values
Update logp Set refid =
  (Select refid From literature_refs Where ref=reference);
-- No need for reference column anymore
Alter Table logp Drop Column ref;
```

This will create a table literature_refs that will hold all the unique values of references that exist in the table `logp`. The comments in the above code should explain the steps in this process. Once the `literature _ refs` table is complete, a full reference can be obtained during a search of `logp` using SQL like this.

```
Select cas, reference From logp Join literature_refs Using (refid);
```

A brief excerpt from a `literature _ refs` table is shown in Table A.1. It was constructed using this technique and nicely illustrates an advantage of normalizing a table in this way.

Table A.1 Sample Rows from Reference
Table Showing How Spelling Anomolies
Can Be Easily Identified

Refid	Reference
2	ABRAHAM MH ET AL. (1994)
3	ABRAHAM,MH ET AL. (1994)
84	CHEM INSPECT TESTING INST (1992)
85	CHEM INSPECT TEST INST (1992)
86	CHEM INSPEXT TEST INST (1992)
131	EL TAYAR,N ET AL. (1985)
132	EL TAYAR,N ET AL. (1991)
133	EL TAYER,N ET AL. (1985)

Spelling anomalies can be easily identified and corrected, assigning the refid of the correct spelling to those that are incorrect. Of course, the original table from which the references came would have to be updated as well to contain the refid for the correct spelling. Concerning spelling and accuracy mistakes, there is little help that a programmatic approach can offer here. A healthy database needs to be well curated. Notice however that putting tables in third normal forms brings such troubles to light and makes correcting them relatively straightforward.

A.4   SQL Functions

Several SQL functions are discussed in the earlier chapters. This section shows the code needed to define these functions and make them available for use in a PostgreSQL database.

A.4.1   Public166keys

This function returns a length 166 fragment key. The input text string ($1 in the function body) is a SMILES, as expected by the matches function. The table of fragments is based on the MACCS 166 public keys[1] and is shown in Table A.5.3 of this Appendix.

```
Create Function public166keys(character varying)
 Returns bit(166) As $EOSQL$
  Select orsum(bit_set(0::bit(166),abit))
  From public166keys Where matches($1,smarts);
 $EOSQL$ Language SQL;
```

A.4.2   Orsum

This is an aggregate function, analogous to the standard SQL aggregate function sum. While sum operates on numeric values, orsum operates on bit strings, or-ing together each input value to provide the result. This example uses the PostgreSQL bit data type and the native function bitor to or together two bit strings.

```
Create Aggregate orsum (
 Basetype = bit,
 Sfunc = bitor,
 Stype = bit
) ;
```

A.4.3   Tanimoto

The tanimoto function is used to compute the Tanimoto similarity of two bitstrings. The input bit strings would have been computed with the public166keys function or another equivalent fragment key or fingerprint function.

```
Create Function tanimoto(bit, bit)
 Returns Real As $EOSQL$
Select nbits_set($1 & $2)::real /
(nbits_set($1) + nbits_set($2) - nbits_set($1 & $2))::real;
$EOSQL$ Language SQL;
```

A.4.4 Euclid

The euclid function is used to compute the Euclidian distance of two bitstrings. The input bit strings would have been computed with the public166keys function or another equivalent fragment key or fingerprint function.

```
Create Function tanimoto(bit, bit)
 Returns Real As $EOSQL$
Select sqrt((nbits_set($1 & $2) + nbits_set(~$1 & ~$2))::real /
 length($1))::real;
 $EOSQL$ Language SQL;
```

A.4.5 Hamming

The hamming function is used to compute the hamming similarity of two bit strings. The input bit strings would have been computed with the public166keys function or another equivalent fragment key or fingerprint function.

```
Create Function tanimoto(bit, bit)
 Returns Real As $EOSQL$
Select ((nbits_set($1 & ~$2) + nbits_set(~$1 & $2))::real /
 length($1))::real;
$EOSQL$ Language SQL;
```

A.4.6 Nbits_set

The nbits _ set function returns the number of bits set to 1 in a bit string. This function is written here using python. Another C language version of this function is shown later in this Appendix.

```
Create Or Replace Function nbits_set(bits bit) Returns Integer
  As $EOPY$
return bits.count('1');
$EOPY$ Language plpythonu Immutable;
```

A.4.7 Amw

This function computes the average molecular weight of an input SMILES structure. It uses the table of atomic weights and SMARTS shown in Table A.2. It relies on the count _ matches function described in Chapter 7.

Table A.2 Average Atomic Weights for Molecular Weight Computation

Smarts	Weight	Symbol
[#1]	1.01	H
[#2]	4.00	He
[#3]	6.94	Li
[#4]	9.01	Be
[#5]	10.81	B
[#6]	12.01	C
[#7]	14.01	N
[#8]	16.00	O
[#9]	19.00	F
[#10]	20.18	Ne
[#11]	22.99	Na
[#12]	24.31	Mg
[#13]	26.98	Al
[#14]	28.09	Si
[#15]	30.97	P
[#16]	32.06	S
[#17]	35.45	Cl
[#18]	39.95	Ar
[#19]	39.10	K
[#20]	40.08	Ca
[#21]	44.96	Sc
[#22]	47.88	Ti
[#23]	50.94	V
[#24]	52.00	Cr
[#25]	54.94	Mn
[#26]	55.85	Fe
[#27]	58.93	Co
[#28]	58.69	Ni
[#29]	63.55	Cu
[#30]	65.39	Zu
[#31]	69.72	Ga
[#32]	72.59	Ge
[#33]	74.92	As
[#34]	78.96	Se
[#35]	79.90	Br
[#36]	83.80	Kr
[#37]	85.47	Rb
[#38]	87.62	Sr

Table A.2 Average Atomic Weights for
Molecular Weight Computation (Continued)

Smarts	Weight	Symbol
[#39]	88.91	Y
[#40]	91.22	Zr
[#41]	92.91	Nb
[#42]	95.94	Mo
[#43]	98.00	Tc
[#44]	101.07	Ru
[#45]	102.91	Rh
[#46]	106.42	Pd
[#47]	107.87	Ag
[#48]	112.41	Cd
[#49]	114.82	In
[#50]	118.71	Sn
[#51]	121.75	Sb
[#52]	127.60	Te
[#53]	126.91	I
[#54]	131.29	Xe
[#55]	132.91	Cs
[#56]	137.34	Ba
[#57]	138.91	La
[#58]	140.12	Ce
[#59]	140.91	Pr
[#60]	144.24	Nd
[#61]	145.00	Pm
[#62]	150.36	Sm
[#63]	151.96	Eu
[#64]	157.25	Gd
[#65]	158.93	Tb
[#66]	162.50	Dy
[#67]	164.93	Ho
[#68]	167.26	Er
[#69]	168.93	Tm
[#70]	173.04	Yb
[#71]	174.97	Lu
[#72]	178.49	Hf
[#73]	180.95	Ta
[#74]	183.85	W
[#75]	186.21	Re
[#76]	190.20	Os

Table A.2 Average Atomic Weights for
Molecular Weight Computation (Continued)

Smarts	Weight	Symbol
[#77]	192.22	Ir
[#78]	195.08	Pt
[#79]	196.97	Au
[#80]	200.59	Hg
[#81]	204.38	Tl
[#82]	207.20	Pb
[#83]	208.98	Bi
[#84]	209.00	Po
[#85]	210.00	At
[#86]	222.00	Rn
[#87]	223.00	Fr
[#88]	226.03	Ra
[#89]	227.03	Ac
[#90]	232.04	Th
[#91]	231.04	Pa
[#92]	238.03	U
[#93]	237.05	Np
[#94]	244.00	Pu
[#95]	243.00	Am
[#96]	247.00	Cm
[#97]	247.00	Bk
[#98]	251.00	Cf
[#99]	252.00	Es
[#100]	257.00	Fm
[#101]	258.00	Md
[#102]	259.00	No
[#103]	260.00	Lr
[*;h1]	1.01	H1
[*;h2]	2.02	H2
[*;h3]	3.03	H3
[*;h4]	4.04	H4
[*;h5]	5.05	H5
[*;h6]	6.06	H6

```
Create Function amw(character varying)
 Returns Numeric As $EOSQL$
    Select sum(weight*count_matches($1,smarts)) From amw;
 $EOSQL$ Language SQL;
```

A.4.8 Tpsa

This function computes the polar surface area of an input SMILES structure. It uses the table for tpsa fragment SMARTS and fragment partial polar surface areas shown in Table A.3. It relies on the count _ matches function described in Chapter 7.

```
Create Function gnova.tpsa(character varying)
 Returns Numeric As $EOSQL$
    Select sum(psa*count_matches($1,smarts)) From tpsa;
$EOSQL$ Language SQL;
```

Table A.3 Fragments Definitions and Partial Surface Areas

psa	Smarts	Description
23.79	[N0;H0;D1;v3]	N#
23.85	[N+0;H1;D1;v3]	[NH]=
26.02	[N+0;H2;D1;v3]	[NH2]-
25.59	[N+1;H2;D1;v4]	[NH2+]=
27.64	[N+1;H3;D1;v4]	[NH3+]-
12.36	[N+0;H0;D2;v3]	=N-
13.6	[N+0;H0;D2;v5]	=N#
12.03	[N+0;H1;D2;v3;!r3]	-[NH]- not in 3-ring
21.94	[N+0;H1;D2;v3;r3]	-[NH]- in 3-ring
4.36	[N+1;H0;D2;v4]	-[N+]#
13.97	[N+1;H1;D2;v4]	-[NH+]=
16.61	[N+1;H2;D2;v4]	-[NH2+]-
12.89	[n+0;H0;D2;v3]	:[n]:
15.79	[n+0;H1;D2;v3]	:[nH]:
3.24	[N+0;H0;D3;v3;!r3]	-N(-)-
3.01	[N+0;H0;D3;v3;r3]	-N(-)- in 3-ring
11.68	[N+0;H0;D3;v5]	-N(=)=
3.01	[N+1;H0;D3;v4]	=[N+](-)-
4.44	[N+1;H1;D3;v4]	-[NH+](-)-
0	[N+1;H0;D4;v4]	-[N+](-)(-)-
17.07	[O+0;H0;D1;v2]	O=
23.06	[O-1;H0;D1;v1]	[O-]-
9.23	[O+0;H0;D2;!r3;v2]	-O- not in 3-ring
12.53	[O+0;H0;D2;r3;v2]	-O- in 3-ring
13.14	[o+0;H0;D2;v2]	:o:
14.14	[n+1;H1;D2;v4]	:[nH+]:
20.23	[O+0;H1;D1]	[OH]-
4.93	[n+0;H0;D3;$(n-*)]	-[n](:):
4.41	[n+0;H0;D3;$(n(:*)(:*):*)]	:[n](:):

Table A.3 Fragments Definitions and Partial Surface Areas (Continued)

psa	Smarts	Description
4.10	[n+1;H0;D3;v4;$(n(:*)(:*):*)]	:[n+](:):
3.88	[n+1;H0;D3;v4;$(n-*)]	-[n+](:):
8.39	[n+0;H0;D3;v5;$(n=*)]	=[n](:):
11.3	[#8+1;H0;D2]	28.5 – 2*8.6
		Nonstandard valency
12.8	[#8+1;H1;D2]	28.5 - 2*8.6 + 1.5
		nonstandard valency
11.3	[#8;H0;D2;v4]	28.5 – 2*8.6
		Nonstandard valency
2.7	[#8;H0;D3;v4]	28.5 - 3 *8.6
		nonstandard valency
2.7	[#8+1;H0;D3;v3]	28.5 – 3*8.6
		Nonstandard valency
7.4	[NH+0;v5;D3]	=N(-)-
		30.5 - 8.2*3 + 1.5
		nonstandard valency
14.10	[#7;v2;D2]	30.5 – 2*8.2
		-[N]-
		Nonstandard valency
15.6	[NH+0;v5;D2]	30.5 - 2*8.2 + 1.5
		-N#, =N=
		nonstandard valency
35.00	[NH3]	30.5 + 3*1.5
		Nonstandard valency?
2.68	O=[N+][O-]	17.07 - 23.06 - 3.01 + 11.68
		(O=) - ([O-])- - (-N(-)-) + (N(=)=)
		fix to make charge-separated spelling work
33.03	N=[N+]=[N-]	23.79 + 13.6 – 4.36
		(N#) + (=N#) - (-[N+]#)
		fix to make charged-separated spelling work.

A.5   Tables Used in Functions

Many of the examples in this book and the functions in this Appendix rely on tables of data to operate. This technique of storing data separately from the function definition makes modification of the data very simple. It also uses all of the data integrity features of a relational database. Data in these tables can be used in various ways, not only in the functions for which they were intended.

A.5.1   Amw

Table A.2 is a table of average molecular weights for each of the first 103 atoms in the periodic table. Because most SMILES do not contain explicit hydrogen atoms, an additional 6 rows are included to match atoms with 1–6 implicit hydrogen atoms. The function amw defined above uses this table to compute average molecular weight.

A.5.2   Tpsa

Table A.3 shows SMARTS for fragments described by Ertl, Rhode, and Selzer.[2] It contains the SMARTS definition of the fragment and the fragment partial polar surface area. This table is used in the tpsa function to compute the polar surface area for a molecular structure.

A.5.3   Public166keys

Table A.4 shows commonly used fragment keys: the MACCS public166-keys. This table is used with the public166keys function above to produce a bit string key for use in filtering before substructure searching and for similarity computations. The table consists of SMARTS patterns[3] used to identify each of 166 substructures.

Table A.4 SMARTS and Bit Position for public166keys

SMARTS	abit	Description
[!0*]	1	ISOTOPE
[Ge,Sn,Pb,As,Sb,Bi,Se,Te,Po]	3	GROUPIVA,VA, VIA PERIODS 4-6
[Ac,Th,Pa,U,Np,Pu,Am,Cm,Bk,Cf,Es,Fm, Md,No,Lr]	4	ACTINIDE
[Sc,Y,Ti,Zr,Hf]	5	GROUPIIIB,IVB
[La,Ce,Pr,Nd,Pm,Sm,Eu,Gd,Tb,Dy,Ho,Er, Tm,Yb,Lu]	6	LANTHANIDE
[V,Nb,Ta,Cr,Mo,W,Mn,Tc,Re]	7	GROUPVB,VIB,VIIB
[!#6]~1~*~*~*~1	8	QAAA@1
[Fe,Co,Ni,Ru,Rh,Pd,Os,Ir,Pt]	9	GROUPVIII
[Be,Mg,Ca,Sr,Ba,Ra]	10	GROUPIIA
~1~~*~*~1	11	4MRING
[Cu,Ag,Au,Zn,Cd,Hg]	12	GROUPIB,IIB

Table A.4 SMARTS and Bit Position for public166keys (Continued)

SMARTS	abit	Description
[#8]~[#7](~[#6])~[#6]	13	ON(C)C
S-S	14	S-S
[#8]~[#6](~[#8])~[#8]	15	OC(O)O
[!#6]~1~*~*~1	16	QAA@1
C#C	17	CTC
[B,Al,Ga,In,Tl]	18	GROUPIIIA
~1~~*~*~*~*~*~1	19	7MRING
[Si]	20	Si
C=C(~[!#6])~[!#6]	21	C=C(Q)Q
~1~~*~1	22	3MRING
[#7]~[#6](~[#8])~[#8]	23	NC(O)O
[#7]-[#8]	24	N-O
[#7]~[#6](~[#7])~[#7]	25	NC(N)N
[#6]@=[#6](@*)@*	26	C$=C($A)$A
I	27	I
[!#6]~[CH2]~[!#6]	28	QCH2Q
[#15]	29	P
[#6]~[!#6](~[#6])(~[#6])~*	30	CQ(C)(C)A
[!#6]~[F,Cl,Br,I]	31	QX
[#6]~[#16]~[#7]	32	CSN
[#7]~[#16]	33	NS
[CH2]=*	34	CH2=A
[Li,Na,K,Rb,Cs,Fr]	35	GROUPIA
[#16&R]	36	SHETEROCYCLE
[#7]~[#6](~[#8])~[#7]	37	NC(O)N
[#7]~[#6](~[#6])~[#7]	38	NC(C)N
[#8]~[#16](~[#8])~[#8]	39	OS(O)O
[#16]-[#8]	40	S-O
C#N	41	CTN
F	42	F
[!#6&!H0]~*~[!#6&!H0]	43	QHAQH
[!#6!#7!#8!#15!#16!#9!#17!#35]	44	OTHER
C=C~[#7]	45	C=CN
Br	46	BR
[#16]~*~[#7]	47	SAN
[#8]~[!#6](~[#8])~[#8]	48	OQ(O)O
[!+0]	49	CHARGE
C=C(~[#6])~[#6]	50	C=C(C)C
[#6]~[#16]~[#8]	51	CSO

Table A.4 SMARTS and Bit Position for public166keys (Continued)

SMARTS	abit	Description
[#7]~[#7]	52	NN
[!#6&!H0]~*~*~*~[!#6&!H0]	53	QHAAAQH
[!#6&!H0]~*~*~[!#6&!H0]	54	QHAAQH
[#8]~[#16]~[#8]	55	OSO
[#8]~[#7](~[#8])~[#6]	56	ON(O)C
[#8R]	57	OHETEROCYCLE
[!#6]~[#16]~[!#6]	58	QSQ
[#16]!:*:*	59	Snot%A%A
S=O	60	S=O
~[#16](~)~*	61	AS(A)A
@!@*@*	62	A$A!A$A
N=O	63	N=O
@!@[#16]	64	A$A!S
c:n	65	C%N
[#6]~[#6](~[#6])(~[#6])~*	66	CC(C)(C)A
[!#6]~[#16]	67	QS
[!#6!H0]~[!#6!H0]	68	QHQH
[!#6]~[!#6!H0]	69	QQH
[!#6!H0]~[!#7]~[!#6!H0]	70	QNQ
[#7]~[#8]	71	NO
[#8]~*~*~[#8]	72	OAAO
S=*	73	S=A
[CH3]~*~[CH3]	74	CH3ACH3
!@[#7]@	75	A!N$A
C=C(~*)~*	76	C=C(A)A
[#7]~*~[#7]	77	NAN
C=N	78	C=N
[#7]~*~*~[#7]	79	NAAN
[#7]~*~*~*~[#7]	80	NAAAN
[#16]~*(~*)~*	81	SA(A)A
*~[CH2]~[!#6!H0]	82	ACH2QH
[!#6]~1~*~*~*~*~1	83	QAAAA@1
[NH2]	84	NH2
[#6]~[#7](~[#6])~[#6]	85	CN(C)C
[CH2]~[!#6]~[CH2]	86	CH2QCH2
[F,Cl,Br,I]!@*@*	87	X!A$A
[#16]	88	S
[#8]~*~*~*~[#8]	89	OAAAO
[!#6!H0]~*~*~[CH2]~*	90	QHAACH2A

Table A.4 SMARTS and Bit Position for public166keys (Continued)

SMARTS	abit	Description
[!#6!H0]~*~*~*~[CH2]~*	91	QHAAACH2A
[#8]~[#6](~[#7])~[#6]	92	OC(N)C
[!#6]~[CH3]	93	QCH3
[!#6]~[#7]	94	QN
[#7]~*~*~[#8]	95	NAAO
~1~~*~*~*~*1	96	5MRING
[#7]~*~*~*~[#8]	97	NAAAO
[!#6]1~*~*~*~*~*~1	98	QAAAAA@1
C=C	99	C=C
*~[CH2]~[#7]	100	ACH2N
[R!r3!r4!r5!r6!r7]	101	8MRINGORLARGER
[!#6]~[#8]	102	QO
Cl	103	CL
[!#6!H0]~*~[CH2]~*	104	QHACH2A
@(@*)@*	105	A$A($A)$A
[!#6]~*(~[!#6])~[!#6]	106	QA(Q)Q
[F,Cl,Br,I]~*(~*)~*	107	XA(A)A
[CH3]~*~*~*~[CH2]~*	108	CH3AAACH2A
*~[CH2]~[#8]	109	ACH2O
[#7]~[#6]~[#8]	110	NCO
[#7]~*~[CH2]~*	111	NACH2A
~(~*)(~*)~*	112	AA(A)(A)A
[#8]!:*:*	113	Onot%A%A
[CH3]~[CH2]~*	114	CH3CH2A
[CH3]~*~[CH2]~*	115	CH3ACH2A
[CH3]~*~*~[CH2]~*	116	CH3AACH2A
[#7]~*~[#8]	117	NAO
~[CH2]~[CH2]~.*~[CH2]~[CH2]~*	118	ACH2CH2A>1
N=*	119	N=A
[!#6R].[!#6R]	120	HETEROCYLICATOM>1
[#7R]	121	NHETEROCYCLE
~[#7](~)~*	122	AN(A)A
[#8]~[#6]~[#8]	123	OCO
[!#6]~[!#6]	124	QQ
!@[#8]!@	126	A!O!A
@!@[#8].*@*!@[#8]	127	A$A!O>1(&..)
~[CH2]~~*~*~[CH2]~*	128	ACH2AAACH2A
~[CH2]~~*~[CH2]~*	129	ACH2AACH2A

Table A.4 SMARTS and Bit Position for public166keys (Continued)

SMARTS	abit	Description
[!#6]~[!#6].[!#6]~[!#6]	130	QQ>1
[!#6!H0].[!#6!H0]	131	QH>1
[#8]~*~[CH2]~*	132	OACH2A
@!@[#7]	133	A$A!N
[F,Cl,Br,I]	134	X(HALOGEN)
[#7]!:*:*	135	Nnot%A%A
O=*.O=*	136	O=A>1
[!#6R]	137	HETEROCYCLE
[!#6][#6H2]*.[!#6][#6H2]*	138	QCH2A>1
[OH]	139	OH
[#8].[#8].[#8].[#8]	140	O>3
[CH3].[CH3].[CH3]	141	CH3>2
[#7].[#7]	142	N>1
@!@[#8]	143	A$A!O
!::*!:*	144	Anot%A%Anot%A
~1~~*~*~*~*1.*~1~*~*~*~*~*1	145	6MRING>1
[#8].[#8].[#8]	146	O>2
~[#6H2]~[#6H2]~	147	ACH2CH2A
~[!#6](~)~*	148	AQ(A)A
[CH3].[CH3]	149	CH3>1
!@@*!@*	150	A!A$A!A
[#7H]	151	NH
[#8]~[#6](~[#6])~[#6]	152	OC(C)C
[!#6][#6H2]*	153	QCH2A
[#6]=[#8]	154	C=O
~!@[CH2&!R]~!@	155	A!CH2!A
[#7]~*(~*)~*	156	NA(A)A
[#6]-[#8]	157	C-O
[#6]-[#7]	158	C-N
[#8].[#8]	159	O>1
[CH3]	160	CH3
[#7]	161	N
a	162	AROM
~1~~*~*~*~*1	163	6MRING
[#8]	164	O
[R]	165	RING
(*).(*)	166	FRAGMENT

A.6 Core Function Implementation for PostgreSQL

Some of the functions described in this book are discussed in an abstract way as chemical extensions to an RDBMS. There is a commercial implementation of these and other functions in the CHORD product of gNova. This section shows three open-source implementations of the core functions. It is necessary to install these additional modules and brief directions are supplied here. It is also necessary to add the plperl and/or plpython procedural languages to the PostgreSQL RDBMS. These are not installed by default, but this is easily done using the `createlang` linux command or the `create language` SQL command. The `createlang` command is part of the installation packages for PostgreSQL.

```
createlang plperl
createlang plpythonu
```

A.6.1 PerlMol/plperlu

PerlMol is a module add-on to the perl language that facilitates working with molecular structures using SMILES, SMARTS, and molfiles, as well as other functionality. PerlMol is available from CPAN, the Comprehensive Perl Archive Network. In order to install PerlMol, it is recommended to use the command `cpan -i PerlMol` as superuser in order to install the modules into the system perl library. This will install all the necessary modules for the following functions, as well as other parts of PerlMol that may be useful.

The following code will define the core functions described in Chapter 7 of this book. The isosmiles function is not included here because of limitation of PerlMol. These functions apply only to the PostgreSQL RDBMS.

```
Create Or Replace Function valid(text) Returns Boolean As $EOPERL$
use Chemistry::File ':auto';

#-- return true if input smi can be parsed
my ($smi) = @_;
my $mol = Chemistry::Mol->parse($smi, format => 'smiles', fatal => 0);

if ($mol->atoms(1)) {
#-- $mol has at least one valid atom
 return true;
} else {
 return false;
}

$EOPERL$ Language plperlu;
```

```
Create Or Replace Function cansmiles(text) Returns Text As $EOPERL$
use Chemistry::File ':auto';

#-- return canonicalized version of input smi
my ($smi) = @_;
my $mol = Chemistry::Mol->parse($smi, format => 'smiles');

return $mol->sprintf('%S');
$EOPERL$ Language plperlu;

Create Or Replace Function keksmiles(text) Returns Text As $EOPERL$
use Chemistry::File ':auto';

#-- return kekulized version of input smi
my ($smi) = @_;
my $mol = Chemistry::Mol->parse($smi, format => 'smiles',
  kekulize => 1);

return $mol->sprintf('%s');
$EOPERL$ Language plperlu;


Create Or Replace Function smiles_to_molfile(text) Returns Text
  As $EOPERL$
use Chemistry::File ':auto';
#--use Chemistry::3DBuilder qw(build_3d);

#-- convert smi to molfile format
my ($smi) = @_;
my $mol = Chemistry::Mol->parse($smi, format => smiles);

#-- compute 3D coords
#--build_3d($mol);
return $mol->print(format => sdf);
$EOPERL$ Language plperlu;

Create Or Replace Function molfile_to_smiles(text) Returns Text
  As $EOPERL$
use Chemistry::File ':auto';

my ($smi) = @_;
my $mol = Chemistry::Mol->parse($smi, format => sdf);

return $mol->print(format => smiles);
$EOPERL$ Language plperlu;

Create Or Replace Function matches(text, text) Returns Boolean
  As $EOPERL$
use Chemistry::File::SMILES;
use Chemistry::File::SMARTS;
use Chemistry::Ring 'aromatize_mol';
```

```
#-- return true if smi is matched by sma
my ($smi, $sma) = @_;
my $mol  = Chemistry::Mol->parse($smi, format => smiles);
aromatize_mol($mol);
my $patt = Chemistry::Pattern->parse($sma, format => smarts);

if ( $patt->match($mol) ) {
 return 't';
} else {
 return 'f';
}
$EOPERL$ Language plperlu;

Create Or Replace Function count_matches(text, text) Returns
  Integer As $EOPERL$
use Chemistry::File::SMILES;
use Chemistry::File::SMARTS;
use Chemistry::Ring 'aromatize_mol';

#-- return how many times smi is matched by sma
my ($smi, $sma) = @_;
my $mol  = Chemistry::Mol->parse($smi, format => smiles);
aromatize_mol($mol);
my $patt = Chemistry::Pattern->parse($sma, format => smarts);

my $nmatch = 0;
while ( $patt->match($mol) ) {
 ++$nmatch;
}
return $nmatch;
$EOPERL$ Language plperlu;

Create Or Replace Function list_matches(text, text, integer)
  Returns Text[] As $EOPERL$
use Chemistry::File::SMILES;
use Chemistry::File::SMARTS;
use Chemistry::Ring 'aromatize_mol';

#-- return list of atoms in smi matched by sma
#-- return nshow'th match; all matches if nshow=0
my ($smi, $sma, $nshow) = @_;
my $mol  = Chemistry::Mol->parse($smi, format => smiles);
aromatize_mol($mol);
my $patt = Chemistry::Pattern->parse($sma, format => smarts);

#---- map atom ids (returned from $patt->atom_map below) to atom
  numbers in mol
my %atom_number;
my $iatom = 1;
foreach ( $mol->atoms ) {
 $atom_number{$_} = $iatom;
 ++$iatom;
}
```

```perl
my @match_list = ();
my @matches = ();
my $nmatch = 0;
while ( $patt->match($mol) ) {
        $nmatches = @matches = $patt->atom_map;
        for (my $i=0; $i<$nmatches; ++$i) {
                $matches[$i] = $atom_number{$matches[$i]};
        }
#-- braces make this a postgresql array
        $match_list[$nmatch] .= "{" . (join ",", @matches) . "}";
        ++$nmatch;
}
return undef unless ($nmatch > 0);

if ($nshow > 0 && $nshow <= $nmatch) {
 return "$match_list[$nshow-1]";
} else {
#-- braces make this a postgresql array (of arrays)
 return "{" . (join ",", @match_list) . "}";
}
$EOPERL$ Language plperlu;

Create Or Replace Function list_matches(text, text) Returns Text[]
  As $EOSQL$
-- Convenience function to return first match
 Select list_matches($1, $2, 1);
$EOSQL$ Language SQL;
```

In order to install these functions into a database named mydb, the follow-ing command would be used. Assume that the code above is stored in a file named perlmol-core.sql.

```
sudo -u postgres psql mydb <perlmol-core.sql
```

It is necessary to install these as the PostgreSQL superuser, here postgres, or as any another PostgreSQL superuser. This is because the "untrusted" language plperlu is used. The language plperlu must be used because of the perl use statements in these functions. It might be more accurate to say "unrestricted" language, since plperlu can use all the functionality of perl. More information about the differences between plperl and plperlu is available.[4]

A.6.2   FROWNS/plpythonu

FROWNS[5] is an open source python module loosely based on Andrew Dalke's PyDaylight.[6] It includes methods that operate on SMILES, SMARTS, and molfiles as well as other functionality, including finger-prints. Once FROWNS is installed, it can be used within any python pro-gram. Using the plpythonu procedural language available in PostgreSQL,

FROWNS python modules are used to show a second way in which the core functions might be implemented. The following plpythonu code extends PostgreSQL with most of the core functions described in Chapter 7. The isosmiles and keksmiles functions are not included here because of limitations of FROWNS.

```
Create Schema frowns;
Grant All On Schema frowns to public;

Create Or Replace Function frowns.valid(smi Text) Returns Boolean
As $EOPY$
from frowns import Smiles
try:
  mol = Smiles.smilin(smi)
  return True
except:
  return False
$EOPY$ Language plpythonu Immutable;

Create Or Replace Function frowns.cansmiles(smi Text) Returns Text
As $EOPY$
from frowns import Smiles
mol = Smiles.smilin(smi)
return mol.cansmiles()
$EOPY$ Language plpythonu Immutable;

Create Or Replace Function frowns.smiles_to_symbols(smi Text)
Returns Text[] As $EOPY$
from frowns import Smiles
mol = Smiles.smilin(smi)
return "{" + ",".join((a.symbol for a in mol.atoms)) + "}"
$EOPY$ Language plpythonu Immutable;

Create Or Replace Function frowns.smiles_to_bonds(smi Text) Returns
Integer[][2] As $EOPY$
from frowns import Smiles
mol = Smiles.smilin(smi)
iatom = 0
return "{" + ",".join( \
  [ "{" + ",".join((str(b.atoms[0].index+1),str(b.atoms[1].
index+1),str(b.bondtype))) + "}" for b in mol.bonds ] \
  ) + "}"
$EOPY$ Language plpythonu Immutable;

Create Or Replace Function frowns.smiles_to_molfile(smi Text, name
Text, coords Numeric[][]) Returns Text As $EOPY$
from frowns import Smiles
# just get the molfile format right.
# stereochemistry missing; charges only appear in CHG records
mol = Smiles.smilin(smi)
(x,y,z) = (0.0, 0.0, 0.0)
if name is None:
```

```
 tname = ""
else:
 tname = name
if coords is not None:
 tcoords = eval( (coords.replace('{','[')).replace('}',']') )
molfile = [];
molfile.append(tname)
molfile.append(' gNova FROWNS smiles_to_molfile')
molfile.append(smi)
atoms = mol.atoms
bonds = mol.bonds
molfile.append("%3d%3d%3d%3d%3d%3d%3d%3d%3d%3d%6d V%4d" % (len(atoms),
len(bonds), 0, 0, 0, 0, 0, 0, 0, 999, 2000))

# oddball mapping of charges
#cmap[-3] = 7
#cmap[-2] = 6
#cmap[-1] = 5
#cmap[0] = 0
#cmap[1] = 3
#cmap[2] = 2
#cmap[3] = 1
cmap = [7,6,5,0,3,2,1]
for a in atoms:
  if coords is not None:
    (x,y,z) = (tcoords[a.index][0], tcoords[a.index][1], tcoords[a.
index][2])
  molfile.append(" %9.4f %9.4f %9.4f %-2s%3d%3d%3d%3d%3d%3d%3d%3d%3
d%3d%3d%3d" % (x,y,z, a.symbol, 0,cmap[3+a.
charge],0,0,0,0,0,0,0,0,0,0))

for b in bonds:
  molfile.append("%3d%3d%3d%3d%3d%3d%3d" % (b.atoms[0].index+1,
b.atoms[1].index+1, b.bondtype,0,0,0,0))

for a in atoms:
  if a.charge:
    molfile.append("M  CHG%3d %3d %3d" % (1, a.index+1, a.charge))
molfile.append("M  END")
molfile.append("$$$$")
return "\n".join(molfile)
$EOPY$ Language plpythonu;

Create Or Replace Function frowns.smiles_to_molfile(smi Text, name
Text) Returns Text As $EOSQL$
 Select frowns.smiles_to_molfile($1, $2, null);
$EOSQL$ Language SQL;

Create Or Replace Function frowns.smiles_to_molfile(smi Text)
Returns Text As $EOSQL$
 Select frowns.smiles_to_molfile($1, null, null);
$EOSQL$ Language SQL;
```

```
Create Or Replace Function frowns.molfile_to_smiles(molfile Text)
Returns Text As $EOPY$
from frowns import MDL
from frowns import Smiles
from frowns.mdl_parsers import Generator
import StringIO
import sys
#fd = StringIO.StringIO(molfile + "\n$$$$")
fd = StringIO.StringIO(molfile)
reader = MDL.sdin(fd)
mol, text, error = reader.next()
fd.close()
if not mol:
  print "Error parsing molfile"
  print error + text
  return None
else:
 # sometimes mol.cansmiles emits [C] when C is proper, this seems
to fix that
  return frowns.Generator.INDEX
  #return (Smiles.smilin(mol.cansmiles())).cansmiles()

#raise ValueError("Error parsing molfile")
return None
$EOPY$ Language plpythonu Immutable;

Drop Type frowns.named_property Cascade;
Create Type frowns.named_property As (name Text, value Text);
Create Or Replace Function frowns.molfile_properties(molfile Text)
Returns Setof frowns.named_property As $EOPY$
from frowns import MDL
import StringIO
fd = StringIO.StringIO(molfile + "\n$$$$")
for mol, text, error in MDL.sdin(fd):
  if not mol:
    print error + text
    return None
  else:
    #mol.fields["cansmiles"] = mol.cansmiles()
    return mol.fields.items()

print "Error parsing molfile"
#raise ValueError("Error parsing molfile")
return None
$EOPY$ Language plpythonu Immutable;

Create Or Replace Function frowns.matches(smi Text, sma Text)
Returns Boolean As $EOPY$
from frowns import Smiles
from frowns import Smarts
mol = Smiles.smilin(smi)
pat = Smarts.compile(sma)
match = pat.match(mol)
```

```
try:
  assert match
  return True
except:
  return False
$EOPY$ Language plpythonu Immutable;

Create Or Replace Function frowns.count_matches(smi Text, sma Text)
Returns Integer As $EOPY$
from frowns import Smiles
from frowns import Smarts
mol = Smiles.smilin(smi)
pat = Smarts.compile(sma)
match = pat.match(mol)
try:
  assert match
  imatch = 0
  for path in match:
    imatch += 1
  return imatch
except:
  return 0
$EOPY$ Language plpythonu Immutable;

Create Or Replace Function frowns.list_matches(smi Text, sma Text,
imatch Integer, istart Integer) Returns Integer[] As $EOPY$
from frowns import Smiles
from frowns import Smarts
mol = Smiles.smilin(smi)
pat = Smarts.compile(sma)
try:
  match = pat.match(mol)
  assert match
except:
  return '{null}'
all_matches = list()
nmatch = 0
for path in match:
  nmatch += 1
  matches = [a.index+1 for a in path.atoms]
  pgarray = "{" + ",".join([str(i-1+istart) for i in matches]) + "}"
  if (nmatch == imatch):
    return pgarray
  all_matches.append(pgarray)

return "{"+ ",".join(all_matches)+ "}"
$EOPY$ Language plpythonu Immutable;

Create Or Replace Function frowns.list_matches(Text, Text) Returns
Integer[] As $EOSQL$
 Select frowns.list_matches($1, $2, 1, 0);
$EOSQL$ Language SQL Immutable;
```

```
Create Or Replace Function frowns.list_matches(Text, Text, Integer)
Returns Integer[] As $EOSQL$
 Select frowns.list_matches($1, $2, $3, 0);
$EOSQL$ Language SQL Immutable;

Create Or Replace Function frowns.fp(smi Text, nbits Integer,
maxpath Integer) Returns Bit As $EOPY$
from frowns import Fingerprint
from frowns import Smiles
mol = Smiles.smilin(smi)
numints = nbits / 32
fp = Fingerprint.generateFingerprint(mol, numInts=numints,
pathLength=maxpath)
return "".join([str(bit) for bit in fp.to_list()])
$EOPY$ Language plpythonu Immutable;

Create Or Replace Function frowns.fp(smi Text) Returns Bit As
$EOSQL$
 Select frowns.fp($1, 512, 7);
$EOSQL$ Language SQL Immutable;

Create or Replace FUNCTION frowns.contains(Bit, Bit) Returns
Boolean As  $EOSQL$
 Select $2 = ($1 & $2);
$EOSQL$ Language SQL Immutable;
Comment On FUNCTION frowns.contains(bit, bit)
 Is 'does first bit string contain all the bits of second';

-- get all info from molfiles in one record for insert into a table
-- see frowns.sql for an example of using frowns.molfile_mol() to
-- insert into a table
Drop Type frowns.mol Cascade;
Create Type frowns.mol As (name Text, cansmiles Text, coords
Numeric[][], atoms Integer[]);
Create Or Replace Function frowns.molfile_mol(molfile Text) Returns
frowns.mol As $EOPY$
from frowns import MDL
from frowns import Smiles
import StringIO
fd = StringIO.StringIO(molfile + "\n$$$$")
for mol, text, error in MDL.sdin(fd):
  if not mol:
    print error + text
    return None
  else:
    # sometimes mol.cansmiles emits [C] when C is proper, amol seems
to fix that
    amol = Smiles.smilin(mol.cansmiles())
    return (mol.name, amol.cansmiles(), \
      "{" + ",".join(["{"+str(a.x)+","+str(a.y)+","+str(a.z)+"}" \
        for a in mol.canonical_list[0][1]]) + "}" , \
      "{" + ",".join([str(a.index+1) \
        for a in mol.canonical_list[0][1]]) + "}" )
```

```
print "Error parsing molfile"
#raise ValueError("Error parsing molfile")
return None
$EOPY$ Language plpythonu Immutable;

Create Or Replace Function frowns.graph(smi Text) Returns Text As
$EOPY$
from frowns import Smiles
try:
 mol = Smiles.smilin(smi)
except:
 return None

hcount = 0
for b in mol.bonds:
  b.bondorder = b.bondtype = 1
  b.aromatic = 0
for a in mol.atoms:
  hcount += a.hcount
  a.aromatic = 0
  a.charge = 0
  nbonds = len(a.bonds)
  if a.valences:
    a.imp_hcount = a.hcount = a.valences[0] - nbonds

return mol.cansmiles() + '.H' + str(hcount)
$EOPY$ Language plpythonu Immutable;
```

Assume that the code above is stored in a file named frowns-core.sql. In order to install these functions into a database named `mydb`, the following linux command would be used.

```
sudo -u postgres psql mydb <frowns-core.sql
```

It is necessary to install these as the PostgreSQL superuser, here postgres, or as any other PostgreSQL superuser. This is because the "untrusted" language plpythonu is used. It might be more accurate to say "unrestricted," since plpythonu can use all the functionality of python.

A.6.3 OpenBabel/python

OpenBabel[7] is program used to interconvert molecular structures from one file format to another. The underlying C++ functions allow operations on SMILES, SMARTS, and molfiles, as well as other functionality including fingerprints. There is a python wrapper for OpenBabel toolkit. Using this module and the plpythonu procedural language in PostgreSQL, the following functions implement the core functions.

```
Create Schema openbabel;
Grant All On Schema openbabel to public;
```

```
Create Or Replace Function openbabel.valid(smi Text) Returns
   Boolean As $EOPY$
import openbabel
try:
  obc = openbabel.OBConversion()
  mol = openbabel.OBMol()
  obc.SetInFormat("smi")
  return obc.ReadString(mol, smi)
  return True
except:
  return False
$EOPY$ Language plpythonu Immutable;

Create Or Replace Function openbabel.cansmiles(smi Text) Returns
   Text As $EOPY$
import openbabel
obc = openbabel.OBConversion()
mol = openbabel.OBMol()
obc.SetInAndOutFormats("smi", "can")
if obc.ReadString(mol, smi):
 mol.SetTitle("")
 for a in openbabel.OBMolAtomIter(mol):
  a.UnsetStereo()
  for b in openbabel.OBMolBondIter(mol):
   b.UnsetWedge()
   b.UnsetHash()
   b.UnsetUp()
   b.UnsetDown()
 return obc.WriteString(mol,1)
else:
 raise ValueError("Error in input smiles")
 return None
$EOPY$ Language plpythonu Immutable;

Create Or Replace Function openbabel.isosmiles(smi Text) Returns
   Text As $EOPY$
import openbabel
obc = openbabel.OBConversion()
mol = openbabel.OBMol()
obc.SetInAndOutFormats("smi", "can")
if obc.ReadString(mol, smi):
 mol.SetTitle("")
 return obc.WriteString(mol,1)
else:
 raise ValueError("Error in input smiles")
 return None
$EOPY$ Language plpythonu Immutable;

Create Or Replace Function openbabel.keksmiles(smi Text) Returns
   Text As $EOPY$
import openbabel
obc = openbabel.OBConversion()
mol = openbabel.OBMol()
obc.SetInAndOutFormats("smi", "smi")
```

```
if obc.ReadString(mol, smi):
 mol.SetTitle("")
 mol.Kekulize()
 return obc.WriteString(mol,1)
else:
 raise ValueError("Error in input smiles")
 return None
$EOPY$ Language plpythonu Immutable;

Create Or Replace Function openbabel.smiles_to_symbols(smi Text)
  Returns Text[] As $EOPY$
import openbabel
obc = openbabel.OBConversion()
mol = openbabel.OBMol()
obc.SetInAndOutFormats("smi", "mol")
if obc.ReadString(mol, smi):
 tbl = openbabel.OBElementTable()
 return "{" + ",".join((tbl.GetSymbol(a.GetAtomicNum()) for a in
  openbabel.OBMolAtomIter(mol))) + "}"

else:
 raise ValueError("Error in input smiles")
 return None
$EOPY$ Language plpythonu Immutable;

Create Or Replace Function openbabel.smiles_to_bonds(smi Text)
  Returns Integer[] As $EOPY$
import openbabel
obc = openbabel.OBConversion()
mol = openbabel.OBMol()
obc.SetInAndOutFormats("smi", "mol")
if obc.ReadString(mol, smi):
 bonds = []
 for b in openbabel.OBMolBondIter(mol):
  if b.IsAromatic():
   bo = 4
  else:
   bo = b.GetBO()
  bonds.append("{%d,%d,%d}" % (b.GetBeginAtomIdx(),
b.GetEndAtomIdx(), bo))
 return "{" + ",".join(bonds) + "}"

else:
 raise ValueError("Error in input smiles")
 return None
$EOPY$ Language plpythonu Immutable;

Create Or Replace Function openbabel.smiles_to_molfile(smi Text)
  Returns Text As $EOPY$
import openbabel
obc = openbabel.OBConversion()
mol = openbabel.OBMol()
obc.SetInAndOutFormats("smi", "mol")
```

```
if obc.ReadString(mol, smi):
 return obc.WriteString(mol)
else:
 raise ValueError("Error in input smiles")
 return None
$EOPY$ Language plpythonu Immutable;


Create Or Replace Function openbabel.molfile_to_smiles(molfil Text)
  Returns Text As $EOPY$
import openbabel
obc = openbabel.OBConversion()
mol = openbabel.OBMol()
obc.SetInAndOutFormats("sdf", "can")
if obc.ReadString(mol, molfil):
 mol.SetTitle("")
 return obc.WriteString(mol,1)
else:
 raise ValueError("Error in input molfile")
 return None
$EOPY$ Language plpythonu Immutable;


Create Or Replace Function openbabel.matches(smi Text, sma Text)
  Returns Boolean As $EOPY$
import openbabel
obc = openbabel.OBConversion()
mol = openbabel.OBMol()
obc.SetInFormat("smi")
if obc.ReadString(mol, smi):

 pat = openbabel.OBSmartsPattern()
 if pat.Init(sma):
  return pat.Match(mol)
 else:
  raise ValueError("Error in input smarts")
  return None

else:
 raise ValueError("Error in input smiles")
 return None
$EOPY$ Language plpythonu Immutable;


Create Or Replace Function openbabel.count_matches(smi Text, sma
  Text) Returns Integer As $EOPY$
import openbabel
obc = openbabel.OBConversion()
mol = openbabel.OBMol()
obc.SetInFormat("smi")
if obc.ReadString(mol, smi):

 pat = openbabel.OBSmartsPattern()
 if pat.Init(sma):
  if pat.Match(mol):
   return len(pat.GetUMapList())
```

```
   else:
    return 0
  else:
   raise ValueError("Error in input smarts")
   return None


 else:
  raise ValueError("Error in input smiles")
  return None
$EOPY$ Language plpythonu Immutable;

Create Or Replace Function openbabel.list_matches(smi Text, sma
  Text, imatch Integer, istart Integer) Returns Integer[] As $EOPY$
import openbabel
obc = openbabel.OBConversion()
mol = openbabel.OBMol()
obc.SetInFormat("smi")
if obc.ReadString(mol, smi):

 pat = openbabel.OBSmartsPattern()
 if pat.Init(sma):
  if pat.Match(mol):
   i = 0
   all_matches = list()
   for p in pat.GetUMapList():
    i += 1
    pgarray = '{' + ','.join([str(a-1+istart) for a in p]) + '}'
    if i == imatch: return pgarray
    all_matches.append(pgarray)
   return '{' + ','.join(all_matches) + '}'
  else:
   return None


 else:
  raise ValueError("Error in input smarts")
  return None


else:
 raise ValueError("Error in input smiles")
 return None
$EOPY$ Language plpythonu Immutable;


Create Or Replace Function openbabel.list_matches(Text, Text)
  Returns Integer[] As $EOSQL$
 Select openbabel.list_matches($1, $2, 1, 0);
$EOSQL$ Language SQL Immutable;


Create Or Replace Function openbabel.list_matches(Text, Text,
  Integer) Returns Integer[] As $EOSQL$
 Select openbabel.list_matches($1, $2, $3, 0);
$EOSQL$ Language SQL Immutable;
```

```
Create Or Replace Function openbabel.fp(smi Text) Returns Bit As
  $EOPY$
import pybel
mol = pybel.readstring("smi", smi)
fp = mol.calcfp()
fpx = ['0' for i in range(1024)]
for b in fp.bits:
 fpx[b-1] = '1'
return  "".join(fpx)
$EOPY$ Language plpythonu Immutable;

Create or Replace FUNCTION openbabel.contains(Bit, Bit) Returns
  Boolean As $EOSQL$
 Select $2 = ($1 & $2);
$EOSQL$ Language SQL Immutable;
Comment On FUNCTION openbabel.contains(bit, bit)
 Is 'does first bit string contain all the bits of second';

-- get all info from molfiles in one record for insert into a table
-- see openbabel.sql for an example of using openbabel.molfile_
  mol() to
-- insert into a table
Drop Type openbabel.mol Cascade;
Create Type openbabel.mol As (name Text, cansmiles Text, coords
  Numeric[][], atoms Integer[]);
Create Or Replace Function openbabel.molfile_mol(molfil Text)
  Returns openbabel.mol As $EOPY$
import openbabel
obc = openbabel.OBConversion()
mol = openbabel.OBMol()
obc.SetInAndOutFormats("sdf", "can")
if obc.ReadString(mol, molfil):
 title = mol.GetTitle()
 mol.SetTitle("")
 cansmi = obc.WriteString(mol,1)
 pat = openbabel.OBSmartsPattern()
 if pat.Init(cansmi):
  if pat.Match(mol):
   map = (pat.GetUMapList())[0]

 return title, cansmi, \
  "{" + ",".join(["{"+str(a.x())+","+str(a.y())+","+str(a.z())+"}" \
    for a in [mol.GetAtom(i) for i in map]]) + "}" , \
  "{" + ",".join([str(i) for i in map]) + "}"

print "Error parsing molfile"
#raise ValueError("Error parsing molfile")
return None
$EOPY$ Language plpythonu Immutable;

Drop Type openbabel.named_property Cascade;
Create Type openbabel.named_property As (name Text, value Text);
```

```
Create Or Replace Function openbabel.molfile_properties(molfil
  Text) Returns Setof openbabel.named_property As $EOPY$
import pybel
mol = pybel.readstring("sdf", molfil)
return mol.data.iteritems()

print "Error parsing molfile"
#raise ValueError("Error parsing molfile")
return None
$EOPY$ Language plpythonu Immutable;

Create Or Replace Function openbabel.graph(smi Text) Returns Text
  As $EOPY$
import openbabel
obc = openbabel.OBConversion()
mol = openbabel.OBMol()
obc.SetInAndOutFormats("smi", "can")
try:
 obc.ReadString(mol, smi)
except:
 return None

hcount = 0
for a in openbabel.OBMolAtomIter(mol):
  hcount += a.ImplicitHydrogenCount() + a.ExplicitHydrogenCount()
for b in openbabel.OBMolBondIter(mol):
  b.SetBondOrder(1)
  b.UnsetAromatic()
for a in openbabel.OBMolAtomIter(mol):
  a.UnsetAromatic()
  a.SetFormalCharge(0)

return obc.WriteString(mol, 1) + '.H' + str(hcount)
$EOPY$ Language plpythonu Immutable;
```

Assume that the code above is stored in a file named openbabel-core.sql. In order to install these functions into a database named mydb, the following linux command would be used.

```
sudo -u postgres psql mydb <openbabel-core.sql
```

It is necessary to install these as the PostgreSQL superuser, here postgres, or as any another PostgreSQL superuser. This is because the "untrusted" language plpythonu is used. It might be more accurate to say "unrestricted," since plpythonu can use all the functionality of python.

A.7   C Language PostgreSQL Functions

C language functions require more work than functions written in other languages. As with all C programs, they must first be compiled. The shared object file output by the complier is then placed in the PostgreSQL library. The following code implements the nbits _ set function that returns the number of bits set in a bit string. This function is also shown in a python version earlier in this Appendix.

```c
#include "postgres.h"              /* general Postgres declarations */
#include "fmgr.h"                  /* for argument/result macros */
#include "executor/executor.h"     /* for GetAttributeByName() */
#include "utils/varbit.h"

/* These prototypes just prevent possible warnings from gcc. */

Datum   nbits_set(PG_FUNCTION_ARGS);
PG_FUNCTION_INFO_V1(nbits_set);
Datum
nbits_set(PG_FUNCTION_ARGS)
{
/* how many bits are set in a bitstring? */

        VarBit     *a = PG_GETARG_VARBIT_P(0);
        int n=0;

        /*
         * VARBITLENTOTAL is the total size of the struct in bytes.
         * VARBITLEN is the size of the string in bits.
         * VARBITBYTES is the size of the string in bytes.
         * VARBITHDRSZ is the total size of the header in bytes.
         * VARBITS is a pointer to the data region of the struct.
         */
        int i;
        unsigned char *ap = VARBITS(a);
        unsigned char aval;
        for (i=0; i < VARBITBYTES(a); ++i) {
                aval = *ap; ++ap;
                if (aval == 0) continue;
                if (aval & 1) ++n;
                if (aval & 2) ++n;
                if (aval & 4) ++n;
                if (aval & 8) ++n;
                if (aval & 16) ++n;
                if (aval & 32) ++n;
                if (aval & 64) ++n;
                if (aval & 128) ++n;
        }
        PG_RETURN_INT32(n);
}
```

Suppose this code exists in a file bits.c. The following linux commands will process this file and make it useable as a new PostgreSQL function.

```
> gcc -shared -o bits.so -I/usr/include/postgresql/8.2/server bits.c
> cp bits.so /usr/lib/postgresql/8.2/lib/bits.so
```

The directories named in these commands are valid for version 8.2 of PostgreSQL running on a Ubuntu linux machine. They may be valid for other distributions of linux, but may need to be adjusted for another installation.

Finally, the function is defined using the following SQL.

```
Create Or Replace Function nbits_set(bit)
 Returns integer AS 'bits', 'nbits_set' Language c Immutable Strict;
Comment On Function nbits_set(bit) Is 'number of bits set';
```

In functions written in other languages, the code appears in the create command, whether it is written in SQL, plpgsql, plperl, or plpython. For C language functions, the name of the shared object, here bits.so, and the name of the c function, here nbits _ set are named in the create command.

A.8   Database Utilities Dbutils

Several utility functions are discussed in Chapter 13 that can be used from the linux command line. These are molgrep, molcat, molview, molarb, molrandom, and molnear. They operate on tables named structure that contain columns of SMILES, fingerprints, and names. Schemas containing tables like these can be created using the smiloader and sdfloader functions described in the next section. This section lists the shell script that defines the molgrep and other commands.

```
#! /bin/bash
# take your pick of options to psql
PSQL='psql -t -A -P fieldsep=,'
PSQL='psql -H'
PSQL='psql -t -A'
# schema where core chemical functions are: gnova,openbabel,frowns,
  perlmol
FUNC="openbabel"
_molgrep() {
$PSQL -c "Select isosmiles,name from \"$2\".structure where $FUNC.
  contains(fp, $FUNC.fp('$3')) and $FUNC.matches(isosmiles, '$3')"
}

_molcat() {
$PSQL -c "Select isosmiles,name from \"$2\".structure"
}

_molview() {
echo "<script src='/marvin/marvin.js'></script>"
if [ "$3" == "" ]; then
 $PSQL -c "Select marvin_view(isosmiles) from \"$2\".structure"
else
 $PSQL -c "Select marvin_view(isosmiles,'$3') from \"$2\".structure"
fi
}
```

```
_molarb() {
$PSQL -c "Select isosmiles,name from \"$2\".structure order by
  md5(id+$4) limit $3"
}


_molrandom() {
$PSQL -c "Select isosmiles,name from \"$2\".structure order by
  md5(id+$RANDOM) limit $3"
}


_molnear() {
$PSQL -c "Select isosmiles,name,tanimoto(fp, $FUNC.fp('$3')) from
  \"$2\".structure where tanimoto(fp, $FUNC.fp('$3')) > $4"
}


_molsame() {
$PSQL -c "select isosmiles,name from \"$2\".structure where
  isosmiles in (select isosmiles from \"$2\".structure intersect
  select isosmiles from \"$3\".structure)"
}


ME=./dbutils # could be /usr/local/bin/dbutils
cmd="$0"
if [ "$cmd" == "bash" ]; then
 alias molgrep="$ME _molgrep"
 alias molcat="$ME _molcat"
 alias molview="$ME _molview"
 alias molrandom="$ME _molrandom"
 alias molarb="$ME _molarb"
 alias molnear="$ME _molnear"
 alias molsame="$ME _molsame"
else
 $1 $*
fi
```

The file containing these shell commands should be created and called
dbutils. In order to define the molgrep and other commands, the command source dbutils is issued. After that, the commands molgrep,
molcat, molview, molarb, molrandom, and molnear become available.
The use of these commands is discussed in Chapter 13.

A.9   Loading Files Into Simple Tables

Most of this book is devoted to explaining how to design schemas to best
suit the needs of a project involving chemical structures. This section is
intended to bridge the gap between using files and using a complex schema
of tables in a relational database. Sometimes it helps to simply get molecular structure files into the database, and decide later how best to integrate
them into new or existing schemas. This section shows two utilities that
create a simple schema that can be used for many purposes, such as those

discussed in Chapter 13. There are two perl scripts shown here: one to load a SMILES file and one to load an sdf file. Each creates a new schema named by the user. Each creates a table named structure containing SMILES, isomeric SMILES, canonical SMILES, names, and fingerprints within that schema. The sdfloader function creates two additional tables named sdf and property. The table sdf contains the minimally processed input file, simply split into separate structures, one per row. The property table contains the names and values of the data items for each structure. The unprocessed text value of each data item is stored as well as the numeric value, if it is possible to convert the text value to a number.

The previous section shows a number of utility functions that operate from the linux command. Those utilities were intended to be used with the tables created using the smiloader and sdfloader scripts shown here. It is also possible to use the data in the tables created by these scripts to create other tables and schemas that are more suited to the needs of a particular project. Any of the other functions described in this book can also be used with these tables.

A.9.1 Smiloader

```perl
#! /usr/bin/perl

$schema = $ARGV[0];
die "Schema name required\nusage: loader schema\n" unless
($schema);

print <<EOSQL;
Drop Schema If Exists $schema Cascade;
Create Schema $schema;
Create Sequence $schema.structure_id_seq;
Create Table $schema.structure (id Integer Primary Key Default
Nextval('$schema.structure_id_seq'), name Text, smiles Text,
isosmiles Text, fp Bit Varying);
Copy $schema.structure (smiles,name) From Stdin;
EOSQL

while (<stdin>) {
 s/\r//; chomp;
 ($smi,$name) = split;
 print "$smi\t$name\n";
}

print <<EOSQL;
\\.
set search_path=openbabel;
Update $schema.structure Set fp=fp(smiles),
isosmiles=isosmiles(smiles) Where valid(smiles);
EOSQL
```

This script can be used as follows.

```
perl smiloader drugs <drugs.smi | psql mydb
```

A.9.2   Sdfloader

```perl
#! /usr/bin/perl

$schema = $ARGV[0];
die "Schema name required\nusage: loader schema\n" unless
($schema);

print <<EOSQL;
Drop Schema If Exists $schema Cascade;
Create Schema $schema;
Create Sequence $schema.structure_id_seq;
Create Table $schema.sdf (id Integer Default Nextval('$schema.
structure_id_seq'), molfile Text);
Create Table $schema.structure (id Integer Primary Key Default
Nextval('$schema.structure_id_seq'), name Text, isosmiles Text,
cansmiles Text, fp Bit Varying, coords Numeric[][3], atoms
Integer[]);
Create Table $schema.property (id Integer References $schema.
structure (id), name Text, tvalue Text, nvalue Numeric);
Copy $schema.sdf (molfile) From Stdin;
EOSQL

while (<stdin>) {
 if (/\$\$\$\$/) {
  print;
 } else {
  s/\r//; chomp; print; print "\\n";
 }
}

print <<EOSQL;
\\.
set search_path=openbabel;
Insert Into $schema.structure (id, name, isosmiles, coords, atoms)
 Select id, (molfile_mol(molfile)).* from $schema.sdf;

Update $schema.structure Set cansmiles=cansmiles(isosmiles) Where
valid(isosmiles);
Update $schema.structure Set fp=fp(cansmiles) Where
valid(cansmiles);

Alter Table $schema.sdf Add Constraint sdf_id_fk Foreign Key (id)
References $schema.structure (id);

Insert into $schema.property (id, name, tvalue)
```

```
select id, (p).name, (p).value from
 (select id, molfile_properties(molfile) as p from $schema.sdf)
atmp;

-- This regexp may not catch all numeric values
Update $schema.property Set nvalue = tvalue::numeric
 Where tvalue ~ E'^[+-]?[0-9]+(\\\\.[0-9]*)?([Ee][+-]?[0-9]+)?\$';
-- You may choose to name colums to be converted to numeric
--Update $schema.property Set nvalue = tvalue::numeric
-- Where Name = 'IC50_uM';
EOSQL
```

This script can be used as follows.

```
perl sdfloader vla4 <vla-4.sdf | psql mydb
```

There are many sources of sample structure files. The examples used in this book come from the pubchem project[8] and the QSAR world project.[9] There are SMILES files available from the National Cancer Institute of the National Institutes of Health.[10]

The drugs.smi file is used in several places in the previous chapters. This file contains the SMILES shown in Table A.5.

Table A.5 SMILES and Names Used in Sample Table Named Drugs

CN1C2CCC1CC(C2)OC(=O)C(CO)c3ccccc3	atropine
c1cc(ccc1C(C(CO)NC(=O)C(Cl)Cl)O)N(=O)=O	chloramphenicol
c1c2c(cc(c1Cl)S(=O)(=O)N)S(=O)(=O)N=CN2	chlorothiazide
CN(C)CCCN1c2ccccc2Sc3c1cc(cc3)Cl	chlorpromazine
Cc1c(nc[nH]1)CSCCNC(=NC#N)NC	cimetidine
CN1c2ccc(cc2C(=NCC1=O)c3ccccc3)Cl	diazepam
CC(=O)OC1C(Sc2ccccc2N(C1=O)CCN(C)C)c3ccc(cc3)OC	diltiazem
CN(C)CCOC(c1ccccc1)c2ccccc2	diphenhydramine
c1ccc(c(c1)C(=O)O)Nc2cccc(c2)C(F)(F)F	flufenamic acid
c1cc(ccc1C(=O)CCCN2CCC(CC2)(c3ccc(cc3)Cl)O)F	haloperidol
CN(C)CCCN1c2ccccc2CCc3c1cccc3	imipramine
CCN(CC)CC(=O)Nc1c(cccc1C)C	lidocaine
CCC1(C(=O)NC(=O)NC1=O)c2ccccc2	phenobarbital
c1ccc(cc1)C2(C(=O)NC(=O)N2)c3ccccc3	phenytoin
CCN(CC)CCNC(=O)c1ccc(cc1)N	procainamide
CC(C)NCC(COc1ccc2c1cccc2)O	propranolol
CCCCNc1ccc(cc1)C(=O)OCCN(C)C	tetracaine
COc1cc(cc(c1OC)OC)Cc2cnc(nc2N)N	trimethoprim
CC(C)C(CCCN(C)CCc1ccc(c(c1)OC)OC)(C#N)c2ccc(c(c2)OC)OC	verapamil

References

1. Durant, J.L., Leland, B.A., Henry, D.R., and Nourse, J.G. 2002. Reoptimization of MDL keys for use in drug discovery. *J. Chem. Inf. Comput. Sci.* 42(6):1273–1280.
2. Ertl, P., Rohde, B., and Selzer, P. 2000. Fast calculation of molecular polar surface area as a sum of fragment-based contributions and its application to the prediction of drug transport properties. *J. Med. Chem.* 43:3714–3717.
3. Yang, J.J. 2006. Private communication.
4. PostgreSQL. Trusted and untrusted PL/Perl. Plperl. http://www.postgresql. org/docs/8.0/interactive/plperl-trusted.html (accessed April 18, 2008).
5. Kelley, B. Frowns ChemoInformatics System. 2002. http://frowns.source-forge.net/ (accessed April 18, 2008).
6. Dalke, A., PyDaylight II. 2000. http://www.daylight.com/meetings/mug00/Dalke/index.html (accessed April 18, 2008).
7. Open Babel: The open source chemistry toolbox. http://openbabel.org/wiki/Main_Page (accessed April 18, 2008).
8. PubChem text search. http://pubchem.ncbi.nlm.nih.gov/ (accessed April 18, 2008).
9. Porter, J. 2003. VLA4 dataset. http://www.qsarworld.com/qsar-datasets-porter.php. (accessed April 18, 2008).
10. National Cancer Institute. Downloadable structure files of NCI open database compounds. 2007. http://cactus.nci.nih.gov/ncidb2/download.html (accessed April 18, 2008).

Index